Mushrooms and Fungi

Other titles in the series

Wild Flowers
Birds
Aquarium Fish
Stones and Minerals

Chatto Nature Guides

British and European

Mushrooms and Fungi

Illustrated and identified with colour photographs

Andreas Neuner

English translation by Fred Bradley
Edited by D. M. Dring

Chatto & Windus · London

Published by
Chatto & Windus Ltd.
40 William IV Street
London WC2N 4DF

*

Clarke, Irwin & Co Ltd
Toronto

British Library Cataloguing in Publication Data

Neuner, Andreas
Mushrooms and fungi. —
(Chatto nature guides).
1. Fungi — Europe — Identification
I. Title II. Dring, Donald Malcolm
589.2'094 QK606.5

ISBN 0-7011-2327-3
ISBN 0-7011-2328-1 Pbk

Printed in Italy

Introduction

I have often been asked to recommend a book on fungi in which *all* fungi are described in detail and illustrated in colour. Unfortunately my answer has to be disappointing: such a book on fungi does not exist, because the number of varieties of fungus (several thousand different higher fungi in Britain alone) is so large that it would have to be a veritable encyclopaedia running into many volumes; its price would be prohibitive, consulting it cumbersome, and the mass of its information discouraging to the reader.

In view of this, demands for such a book become more modest, and a guide that contains good illustrations and descriptions of the most important edible mushrooms and their dangerous counterparts will be found adequate. I have written this book to meet this demand. It contains a selection of 91 species of importance to the collector of fungi, and leaves out the large number of those mushrooms which, because of their small size, their rarity, or because they are of no importance in the kitchen, are of no interest to the majority of collectors.

The book is intended to be an elementary guide to mycology on which those collectors who have graduated from mere consumers to fungus lovers, to whom the question of "edible or poisonous" is no longer the main problem, can base further studies to build up their knowledge of the subject.

If you know 91 species thoroughly you can call yourself a fungus expert, but should remember that even at this stage you are only at the beginning. Those who claim to "know all about fungi" are either unaware of the large number of varieties or simply trying to show off. The person who knows every single species of fungus does not exist.

It should not be too difficult, however, to identify the 91 fungi illustrated here, and it is most important for the collector to become familiar with those fungi that are poisonous, especially with those which cause not only stomach upsets of varying severity, but are in addition potentially lethal.

Every collector of fungi must be able to identify positively the native Death Cap, and Destroying Angel, and on the continent, Tricholoma pardinum, the Panther Cap (Amanita pantherina), and Inocybe patouillardi.

It would be useful to have a foolproof means, such as a test strip used to detect sugar in urine, which will turn brilliant red when dipped into a dish of poisonous fungi. Unfortunately no such indicator exists and methods used in the past for detecting suspect fungi have long been shown up as figments of superstition. It was believed, for instance, that when a silver spoon or an onion boiled with the fungi turned black this was an infallible sign that the fungi were poisonous. Our attitude should be guided neither by exaggerated fear nor by carelessness. We can read in old cookery books for instance, that even the stoneware pot in which the poisonous dish had been prepared must be destroyed because its continued use would produce symptoms of poisoning. The most important principle is surely to collect only those fungi for the table which are positively known. Species that cannot be satisfactorily identified should be used only after an expert has declared them edible. When you add a new fungus to your menu after thorough examination of its edibility, you are advised to test it for its individual compatibility. Do not eat a large portion of a species of tasty edible fungus the first time you prepare it; be satisfied with a smaller portion to begin with for you may be one of those with whom it does not agree, and a type of fungus eaten by others with impunity may upset you. If you can eat the small sample portion without trouble, you will be able to consume larger quantities.

If in spite of all precautions symptoms of poisoning occur after a meal of mushrooms, treat them as an emergency and call a doctor without delay. If the symptoms appear soon after the meal the condition may be relatively harmless and can sometimes be cured instantly with suitable treatment. But if they are delayed until 11-12 hours later they may indicate Death Cap poisoning, which should be taken very seriously indeed. The quantity of mushrooms consumed as well as the patient's general state of health will determine the course of the illness. In this context I must mention a poisonous fungus (not collected in Britain) that causes symptoms which in certain conditions do not appear until a fortnight after consumption, so that their connection with a fungus meal is by no means readily diagnosed. Once again—beware of unknown species.

Most cases of "mushroom poisoning", however, are not caused by the consumption of poisonous, but of rotten fungi. Fungi decay quickly, particularly when kept in plastic bags, and in late autumn if they have been exposed to ground frost. The doctor will want to examine remains of the collected fungi, because this will enable him to diagnose the type of poisoning and to suggest suitable treatment more quickly. Many who have sought clinical treatment for "mushroom poisoning" have come away with the diagnosis not of what they suspected, but of a stomach or an intestinal ulcer. Even the fear, the imagination of having eaten poisonous fungi may produce the feeling of mushroom poisoning. If you have collected so many mushrooms that you do not know what to do with them you may wish to give some of them to friends, or to send out invitations to a mushroom dinner, but only if you are perfectly confident that you know your fungi. Otherwise you may risk, for all your good intentions, prosecution for causing actual bodily harm, if not death by negligence!

Do not let these serious warnings discourage you altogether. A dozen fresh Edible Boletuses will gladden your heart. But even if the yield is small, collecting mushrooms is a rewarding hobby in the bracing, pure woodland air. Several hours of looking for mushrooms with hundreds of "knee bends" are a thoroughly healthy physical exercise. And there is the beauty of the wood if (changing the saying slightly) you can see it for the mushrooms. What you have found yourself will give you more pleasure than what you have bought in the market.

Collecting fungi requires certain *items of equipment.* No optician stocks "mushroom spectacles" with which you will instantly detect even the most cleverly hidden boletus without fail. But practice will considerably improve your powers of observation. The beginner will be astonished when the expert following in his footsteps picks up a Rough-stemmed Boletus here, a Red Boletus there, which his untrained eye had completely overlooked.

A basket is recommended as a suitable container. Plastic bags, as already mentioned, should never be used. In British woods mushroom gathering is less of a hazard than in Italian ones, where a long, stout stick and a serum against poisonous snake bites, available from any chemist, should be carried at all times. The stick is also recommended elsewhere on the continent in forests situated in rabies-infested areas. Poisonous fungi may not be aggressive: rabid foxes certainly are. Rabies is transmitted through the bite of infected animals, but never through a fungus even if a rabid animal had been in contact with it. Even if its saliva had touched the fungus, transmission of rabies is virtually impossible, because the virus responsible is very quickly destroyed by the action of light. No case of rabies infection as a result of the consumption of edible fungi has become known.

The edible fungi gathered should be cleaned and needles, moss, and slugs removed, preferably before you leave the wood. It is best to cut the mushrooms in half with a knife to see whether they have been attacked by maggots. True nature lovers never leave a trail of dead mushrooms behind. I was once asked by the editor of a daily paper to draw a map not only of the woods, but also of the precise areas, marked with appropriate symbols, where edible mushrooms were known to be found. I would not have obliged him even if the compilation of such a "mushroom atlas" were not totally impossible. No mushroom collector in his senses will reveal the places from which he obtains "his" boletuses year after year without fail. On the contrary, he will if possible camouflage these habitats and, if someone approaches them too closely for comfort, casually mention that although he had never found anything at all in this neck of the woods, only today had startled two 3ft. adders, from which he was able to jump away just in the nick of time. He'd rather deal with a poisonous fungus than with a bite by a poisonous snake any time. Asking rival mushroom collectors is therefore not the right way to find rewarding mushroom sources.

You must know where to look for them. In the woods, obviously, although fungi grow also elsewhere, for instance in gardens, parks, meadows, stubble fields, swamps, burnt ground, on trees and timber, in cellars, mines, and even hothouses. Almost all varieties of fungus depend on closely defined habitats (biotopes). Naturally, the fungus flora of faraway countries differs from our own and only a few species, the so-called cosmopolitans, are found throughout the world. Sandy pine woods are particularly rich in fungi, not so much in the number of varieties as in sheer quantities.

Other species thrive in fir woods. Larch woods have a fungus population found nowhere else. Our beech woods, too, are very rich in different species. Riparian forests (along the banks of large rivers) offer a harvest of morels in spring. The experienced collector will make a point of visiting isolated groups of oak or birch trees in a wood; some mushrooms will be found only under these types of tree.

To grow mushrooms in one's own back garden is an often-voiced ambition. But no seedsman stocks "mushroom seeds". All attempts to obtain convenient mushroom crops in this way have so far failed. Only a few species of fungus growing on wood can be cultivated on a suitable substrate.

As important to the fungus lover as the quest for the mushroom eldorado, the treasure trove of the mushroom collector, is the knowledge when to enter this fabulous land of his dreams.

The *mushroom season* proper, when most varieties appear and are also offered in the local markets, lasts from about the middle of July to the middle of October, naturally varying a little from one year to the next. In Switzerland, even March presents a culinary delight, Hygrophorus marzuolus, which is particularly highly prized. April and May are the months during which to look for morels (Morchella) in riparian forests. June is the time of the gorgeous Russulae. The following four months offer a great multitude of species growing on woodland soil. During November, when the first frosts herald the end of the mushroom season, Lyophyllum and Hygrophorus will be found. During the winter months, especially during periods of thaw, experts look for Flammulina velutipes. During January and February, the mushroom enthusiast, who has by now become passionate in the pursuit of his hobby, collects agarics, which do not die in the autumn and often survive for many years.

Many mushroom collectors set out at the crack of dawn lest an early riser beat them to their own boletus grounds. If their suspicion is justified, no objection can be raised against setting out in darkness. But those who know other species of mushroom, as many as 60 perhaps, need not start on their excursion until the others are returning from the woods. The Edible Boletus may be very sought after, but the connoisseur will find a few species in midmorning which are equal or only little inferior to it in taste. Many have nevertheless returned empty-handed from the woods and disappointed even at the height of the season because they had found absolutely nothing. What they did not know was that the growth of fungi is greatly influenced by the prevailing weather. During prolonged periods of drought and heat, when Europe is sweltering under an anticyclone and the weather is ideal for the swimming pool, it is completely futile to look for mushrooms. At such a time it is better to follow the call of the nearest river bank or beach. To be able to grow, mushrooms need a lot of moisture.

A wet summer, with one depression chasing another across Britain, only briefly interrupted by small areas of high pressure, offers the ideal weather for mushroom collectors, even if others are less enthusiastic about it. Good and bad mushroom years follow each other at random. But even a very poor year often makes up for the disappointment with a late harvest in autumn in a damp atmosphere and fog.

Conscientious, nature-loving mushroom collectors never take the whole mushroom out of the soil, but cut off the stem so as not to damage the subterranean mushroom network, the mycelium. This consideration is appreciated. But in cases where the base end of the stem exhibits the vital distinctive features, for instance in the Death Cap and the Destroying Angel (Amanita phalloides and virosa) and in the Yellow-staining Mushroom (Agaricus xanthodermus), cutting off the stem is not altogether recommended. The blanket of moss should never be lifted to any extent in the search for young mushrooms covered by it.

No matter how excellent the illustrations and how detailed the descriptions in a book on fungi, go to the woods if possible in the company of a mushroom expert. You will find one through a mushroom collectors' club or, on the continent, a fungus advice bureau. In Britain, the British Mycological Society and various local Natural History Societies organize collecting expeditions, called Forays. The BMS distributes lists of these to its Associates and Members. For details of Associateship and Membership you should consult the Membership Secretary, Dr. N. J. Dix, Biology Department, The University, Stirling, Scotland.

For a proper understanding of the description of the various species in this book the knowledge of a few technical terms is necessary. Most of the mushrooms shown have a cap (pileus) and a stem (stipe). Important to the identification of a fungus are the properties of the cap, its size, shape, and colour, whether it is smooth or scaly, dry or sticky, convex or funnel-shaped, of felt-like or velvety texture. The condition of the margin of the cap, too, is significant. It may be rounded or sharp, grooved or have fringes. But the underside of the cap is of much greater importance; to examine it turn the cap upside down. The differences are striking. The underside often consists of gills, which places the fungus in the family Agaricaceae. The shape and colour of the gills, the way in which they are joined to the stem, the quality of their edges, are very varied, and essential pointers to the identification of the species.

Other fungi, such as the Edible Boletus, have a dense layer of tubes instead of gills; their family is called Boletaceae. A third family, Hydnaceae by name, displays a large number of short spines on the underside of the cap, making it look like a stubble field in miniature. Stemless (sessile) agarics have pores on the underside of their caps that look like tiny pinpricks; they are classed among the pore fungi and live on trees. Lastly, there are very rarely found pileated mushrooms in which the spore-producing layer (the hymenium) is completely smooth.

The spores can be seen only under a microscope, because they measure no more than a few thousandths of a millimetre in length. But many hundreds of thousands of spores are visible to the naked eye in the form of spore dust. A spore-print is made to examine it. Separate the cap from the stem and place the former with the spore-producing layer face downwards on a sheet of white paper. On the following morning the paper will be covered with a brown, reddish, black, purple, or white dust consisting of an unimaginable number of spores. The colour of the dust is a clue to the species of the fungus producing it. Often it is the same as that of the gills.

Detailed examination of the stem reveals further distinguishing features. Apart from its length and circumference the stem may be slimy or dry, slender or bulbous, hollow or solid, hoary or shiny. Many fungi have a membranous structure surrounding the stem; it is called a ring and represents the remains of a partial veil, which in the young specimen forms a protective cover of the gills or tubes. In the Cortinariaceae a veil of fine hairs which may be difficult to see extends from the margin of the cap to the stem. Whether the stem grows from a membranous sheath is a matter requiring special attention. This sheath (Volva) and remnants of skin on the cap are all that is left of what is called the universal veil, which completely envelops the young mushroom.

Cap and stem form the fruit body (sporophore) of the fungus. The fungus proper, the mycelium, grown from a spore, is a subterranean network of threads which superficially resembles roots of plants.

A special difficulty besetting the correct identification of a fungus consists in the fact that when young, many of them look very different from the fully grown specimen. A young and a fully grown Fly Agaric resemble each other about as much as humans at the ages of three months and of thirty years. A species is known for certain only after it has been observed at all stages of its development from its first appearance to its decay.

It should also be borne in mind that a fungus does not always have the typical appearance illustrated in a guide book, but may assume a different aspect according to the weather and its habitat.

Fungi and trees mutually promote their growth. Some fungi have entered into a close biological relationship (symbiosis) with certain types of tree. Without the larch to sustain it you will find, for instance, no Suillus elegans. If fungi are removed from a part of the wood through excessive gathering, the trees and the soil will suffer. A law was passed in Italy in 1972 applying to certain parts of the country to prevent more than 2kg of fungi, even inedible ones, being collected by any one person in one day. Such legislation may become necessary in other countries too.

Increased leisure and the possibility of reaching woods quickly by car have made fungus collecting a popular hobby. Please do not knock over or trample underfoot fungi you do not know; spare them if they are too young, too small, or too old, maggot-eaten or partially decayed, when they will merely add weight to your basket and end up in the dustbin anyway.

Books on fungi written during the first half of the century were all based on the well-meaning intention to protect the mushroom collector against illness and death from poisonous fungi. But today, protection of the environment and ecology demand that the fungi be saved from the thoughtlessness of irresponsible "nature lovers". The present book is designed as an aid to fungus lovers and to the threatened fungi alike.

Mushrooms and Fungi

The sequence of the species conforms to the scientific classification Key to symbols:

edible

poisonous

study the description, especially the paragraph "Use"

Editor's Note: Turn to p. 142 for fungus diagram.

The Edible Boletus or Penny Bun

Boletus edulis Bull

The Edible Boletus is the prototype, the edible mushroom *par excellence,* with its typical brown cap, the layer of tubes, and the unchanging white flesh. The more a specimen diverges from this appearance, the more suspect it will be to the layman, who may exclude it from the category of edible fungi, sometimes certainly wrongly, if it is red, blue, not to mention green, has gills on the underside of its cap, or its flesh changes colour. The Edible Boletus and the following 15 species belong to the family Boletaceae. This means that the hymenium (the fertile, spore-bearing layer), lines fine tubes on the underside of the cap, in which the spores are produced, not radiating plates (gills). **Characteristics:** Cap: globular and cushion-shaped, at first very light in colour, gradually turning brown and becoming wrinkled. Tubes: in the young specimen almost white, turning yellowish with advancing age, and finally olive green. Stem: tough, club-shaped, grey-brown. Close observation reveals a very delicate, light-brown network. Flesh: white and firm, does not change colour. The parts below the skin of the cap are sometimes of a brownish colour.—**Habitat:** Inhabits deciduous as well as coniferous woods, especially along the edges; found from July to late autumn. Near large cities sometimes almost extinct, but still abundant in good mushroom years.—**Possibilities of confusion:** With Tylopilus felleus (often; see next illustration).—**Use:** The Edible Boletus is a highly-prized, much sought-after mushroom, also popular with animals. Old specimens are usually badly eaten by the maggots of the Mushroom Fly. It lends itself to the most varied ways of preparation for the table.—**Relatives:** Boletuses found at the beginning of July are of the Boletus aestivalis Paulet variety. Very dark-skinned boletuses, with a red-brown network and growing mostly in fir woods, are called Boletus pinicola. Both types are tasty. Other members of this family resembling the Edible Boletus are rare and should not find their way into the kitchen if only for reasons of their conservation.

Tylopilus felleus

Every year some people find that the season's Edible Boletuses tasted bitter, and think this had something to do with the weather, if not with the radioactive pollution of the atmosphere. The answer is: with neither, but that one or more specimens of Tylopilus felleus had been included in the dish of mushrooms by mistake.—**Characteristics:** Tylopilus felleus does indeed look deceptively like an Edible Boletus, especially when young. Cap: same size as that of the Edible Boletus, light brown, of slightly felt-like texture, in dry weather with a network of fine cracks. Tubes: at first white as in the Edible Boletus, later pale pink; they protrude cushion-like from the underside of the cap; but in the Edible Boletus the colour of the tubes changes from white through yellow to greenish. Stem: not as bulbous as that of the typical Edible Boletus, and more slender. It has a striking, and in old specimens almost dark brown network, which is coarse-meshed. Flesh: white, granular, has little scent. A small sample, however, reveals an unpleasantly bitter taste. This applies also to very young specimens, where it is the only feature distinguishing it from the Edible Boletus.—**Habitat:** Tylopilus felleus is fairly common, particularly in coniferous woods during summer and autumn.—**Possibilities of confusion:** Tylopilus felleus, especially when young, can be distinguished from the Edible Boletus only through sampling it; collectors therefore often mistake it for the latter:—**Use:** Tylopilus felleus is not poisonous, but even a single specimen can thoroughly spoil a lovingly prepared dish of Edible Boletuses, making it quite unpalatable. All attempts to neutralize the bitter taste of the fungus have failed so far. The verdict of the expert that the collector's prized "harvest of Edible Boletuses" should be consigned to the dustbin has at times produced tears of disappointment. Every mushroom collector should be able to identify Tylopilus felleus, the double of the Edible Boletus, without fail.—*Note:* All fungi that have a bitter taste are unsuitable for the table.

Boletus luridus

The two species shown, Boletus luridus and the Satan's Fungus or Devil's Boletus, were never clearly distinguished in old books on fungi and often confused with each other. The value of the fungi, previously described as poisonous, for the table was in doubt. The very name of Boletus luridus and the red hues of the fungus together with the strikingly blue colour of the flesh when it was cut aroused suspicion.—**Characteristics:** Cap: at first hemispherical, gradually flattening, reaching an average diameter of 12cm and more. Its colour varies between light and dark brown. Tubes: growing to a length of about 2cm; the pores are bright red, becoming progressively paler with age; when bruised they turn blue. Stem: at first stout, bulbous, later somewhat slimmer. Its distinct, reticular pattern is striking; its basic colour is pale yellow, against which the red-brown mesh of the network stands out clearly. Flesh: yellowish, and usually bright red at the base of the stem. Depending on the water content of the fungus, the flesh when cut turns more or less deep blue, and finally becomes pale grey. When the fungus is cut a red line appears between the pores and the flesh of the cap, but soon disappears.—**Habitat:** A common fungus in Central Europe, less, so in Britain which likes beechwoods and parkland on chalky soil; summer and autumn.—**Possibilities of confusion:** With Boletus erythropus described below.—**Use:** Was previously considered poisonous. This is true when it is eaten raw. But cooked sufficiently it is one of the good edible mushrooms. Only someone allergic will react with intestinal illness.

Boletus erythropus

Closely resembling Boletus luridus and commoner. But even the beginner can easily identify it, because its stem does not display the previously-described network pattern but is covered with delicate, reddish flakes. The previously mentioned red line is absent. Its taste is superior to that of Boletus luridus, and it is often even more highly prized than the Edible Boletus.

The Satan's Mushroom or Devil's Boletus

Boletus satanas

Many mushroom collectors claim to have found the Satan's Mushroom quite often. But if their finds are examined they will almost always turn out to consist of either Boletus luridus or Boletus erythropus (see illustrations on the preceding page); for the Satan's Mushroom is rather rare and absent from many regions.—**Characteristics:** Cap: diameter up to 25cm. A feature very typical of it is its strikingly bright, whitish colour with a slight olive hue. Tubes: yellowish, more olive green when old, turning blue when injured. The pores are yellowish in the young specimen only, and blood red in the fully grown one. As the mushroom grows older they lose their initial brilliance. The tubes can be readily detached from the cap. Stem: very thick and bulbous, and in proportion to the cap strikingly short. A red network on a yellow background is typical. Flesh: the at first granular and firm flesh turns blue when cut, but never as deeply as that of Boletus luridus or Boletus erythropus. It has an unpleasant smell even in young specimens which becomes stronger with age, when it will be quite offensive in its resemblance to carrion.—**Habitat:** The Satan's Mushroom is rare, but most frequently found on chalky soil in beech, and occasionally in other deciduous woods, always in small groups.—**Possibilities of confusion:** With Boletus luridus and Boletus erythropus (frequently), which have the distinguishing feature of their flesh turning a deeper blue when cut.—**Use:** The Satan's Mushroom is a dangerous poisonous fungus which may cause severe illness. But not a single fatal case of poisoning has come to light. The deterring name is nevertheless justified.—**Relatives:** Four species of boletus are closely related and similar to the Satan's Mushroom. They are rare and of little importance to the mushroom collector; Boletus rhodoxanthus and Boletus purpureus are two of them.

The rule of thumb that the Boletaceae family contains no deadly fungus is correct. But it is as well to remember that it includes bitter varieties, which are unsuitable for consumption.

The Cep

Xerocomus badius

Although the Cep and the two following mushrooms, the Goat's Lip (Xerocomus subtomentosus) and Xerocomus chrysenteron also have pores, other features distinguish them clearly from the boletuses proper. Xerocomus are never moist in dry weather, their pores are larger; they are not as fleshy and therefore not as substantial as the boletuses, and their stems are more slender. It is therefore understandable that mycologists recently separated them from the boletuses and gave them a genus of their own, Xerocomus.—**Characteristics:** Cap: average diameter 5—12cm, larger specimens rare. At first hemispherical, flattening later, occasionally developing a depression in the centre. The colour of the cap is brown, similar to that of the seed coat of the edible chestnut. Tubes: reach a length of more than 1cm, gradually change from yellow to green. They are distinctly sinuate (deeper round the stem). Finger pressure on the tubular layer turns it blue. In very dry weather this colour reaction is weak. Stem: length about 8cm, diameter 2cm, often tapering towards the end, colour whitish to light brown, no network pattern. Flesh: white, changes to light blue, granular, pleasant scent and taste when young.—**Habitat:** Very common in coniferous, deciduous, as well as mixed forests. It seems to have a special liking for 60-70 year old spruce, where it grows at the base of the trunk, often between the root swellings.—**Possibilities of confusion:** With Tylopilus felleus (p. 16) when young. When displaying its typical shape, the Cep is easily identified; but it is known for its unusually wide variations, less in colour than in shape, from very slender, with a stem only 5mm in diameter, to the occasional "middle-aged spread", so that the collector is led to assume that it is a different species altogether. A mushroom similar to the Cep, which, however, turns an unusually deep blue very quickly, is the edible Boletus pulverulentus.—**Use:** The Cep is a much sought-after edible mushroom, which in areas where the Edible Boletus has become rare offers an almost equivalent substitute.

The Goat's Lip

Xerocomus subtomentosus.

The Goat's Lip and Xerocomus chrysenteron are members of the same family as the Cep. The beginner will not find it easy to distinguish between the two closely related species.—**Characteristics:** Cap: grows to a diameter of 10cm and, because its skin is like fine felt, feels like chamois leather to the touch. Brown with a hint of olive green. The skin of the cap can be peeled off. Tubes: length up to 2cm, easily detached from the cap. The pores are a brilliant yellow and rather large. Stem: length up to 10cm, diameter up to 2cm, slender, sometimes curved, brownish, never as deep red as that of Xerocomus chrysenteron. Flesh: white, spongy, when cut changes little in colour, at most to light blue; mild taste.—**Habitat:** Common, usually solitary in coniferous as well as in deciduous forests during summer and autumn.—**Possibilities of confusion:** With Xerocomus chrysenteron (often). This is of no practical importance because both are edible.—**Use:** Young Goat's Lips are good edible mushrooms. Older specimens are often infested with mould, when they will be damaging to health.

Xerocomus chrysenteron

Characteristics: Cap: Same colour as, but slimmer than the Goat's Lip, diameter up to 7cm. As a result of dryness the skin of the cap is usually tesselated by cracks. Where slugs have attacked it the flesh can be seen through the skin. Tubes: yellow, but not as brilliant as the Goat's Lip's and tending slightly towards green. Stem: length up to 7cm and more than 1cm in diameter. In typical specimens it is distinctly red. A second generation of this mushroom grows in the autumn; it is somewhat different in appearance, for instance with hardly any red on the stem. Flesh: very soft, at first pale yellow, turning red and blue.—**Use:** Young mushrooms, especially of the autumn generation, are tasty. Old, often mouldy specimens must be left alone. The mushroom is particularly suitable for casserole dishes.—**Relatives:** The Xerocomus mushrooms described here have about half a dozen mostly rare relatives. One of them, Xerocomus parasiticus, is illustrated together with the Common Earth Ball.

The Ringed Boletus

Suillus luteus

The Ringed Boletus and the following four species are tubular fungi which in appearance are clearly distinct from both the boletuses proper and from the genus Xerocomus. They grow only in symbiosis with trees (mycorrhiza), and exclusively with two-needled pines, i.e. the Scotch Pine (Pinus silvestris) and the Mountain Pine (Pinus montanus). They sometimes have a veil, when the stem will be surrounded by a ring. The cap is usually slimy. Even during periods of dry weather remains of pine needles, leaves and blades of grass will be found adhering to the cap.—**Characteristics:** Cap: diameter up to 12cm, at first hemispherical, then spread out, covered by a thick layer of slime. The slimy skin is easily peeled off. Chocolate-brown at first, lighter later. Tubes: butter-yellow at first, turning olive-yellow, with pentagonal or hexagonal pores, adnexed (lightly attached to the stem). Stem: comparatively short, length hardly more than 6cm. At the upper end it has fine, brown spots, thought to be accumulations of spores. A prominent, membranous ring, at first white and gradually turning brown-violet and adhering to the stem is a striking feature of this fungus. Flesh: very soft, with a pleasant, slightly sour taste. It is often severely attacked by the maggots of the Mushroom Fly; lemon-coloured above the tubes.—**Habitat:** Common, especially in pine woods from summer to autumn.—**Possibilities of confusion:** Distinguishing features: the ring is uncommon in tubular fungi; the cover of the cap is extremely slimy; the flesh is yellowish. No slimy tubular fungi are poisonous or dangerous.—**Use:** The Ringed Boletus is one of the tasty edible mushrooms. It decays very rapidly and must therefore be consumed without delay. The mushroom is best cleaned as soon as it is found; simply peel the sticky skin off. There is an additional advantage in doing this: other mushrooms in the basket will not stick to the Ringed Boletuses and become smeary, slippery, and unappetizing. In America the mushroom, widely distributed in pine plantations, is known as "Slippery Jack".

Suillus grevillei (elegans)

Characteristics: Cap: golden-yellow, sticky, skin comes off readily. Tubes: at first yellow, turning slightly grey, decurrent (running down along the stem). Stem: length up to 12cm, solid flesh, slightly flexible; margin of the cap and stem are joined by a white, membranous veil in the young specimen; it tears as the cap opens up, leaving a conspicuous ring round the stem. Flesh: white to yellowish, fairly soft, becoming vaguely discoloured, with a pleasant scent and taste.—**Habitat:** It associates with the larch (Larix decidua) and is always found together with it. It therefore inhabits mainly mountain forests especially in the Alps, but appears also wherever the larch has been planted in conjunction with afforestation schemes, and even under solitary trees.—**Possibilities of confusion:** With Suillus tridentinus, also growing under the larch, similar to Suillus grevillei, and edible.—**Use:** An excellent edible mushroom, goes very well with a roast joint.

Suillus granulatus

Suillus granulatus, too, is a slimy tubular fungus, but has no ring. It derives its name from the fact that the upper part of its stem has tiny, yellow to brown warts.—**Characteristics:** Cap: in the young specimen hemispherical, becoming flat later and growing to a diameter of up to 12cm. At first the margin is slightly involute (curling inwards). The colour changes from red to brown to yellow-brown. The cap is covered with a very slimy skin, which is easily peeled off. Tubes: yellowish, later yellow-brown, the pores are rounded, of little more than 1mm diameter. The numerous milky droplets exuded by the tubular layer are a striking distinguishing feature. Stem: firm flesh, yellowish, length up to 6cm, always without ring. Flesh: white, gradually turning lemon yellow, does not change colour when cut, has a pleasant scent and a slightly sour taste.—**Habitat:** Very common, always associated with pines, summer to autumn.—**Possibilities of confusion:** Look for warts on the stem and droplets of milky fluid.—**Use:** an excellent edible mushroom.

Suillus variegatus

Although Suillus variegatus, too, is a slimy tubular fungus, it displays its typical properties during wet weather only. When dry it resembles a Xerocomus more closely.—**Characteristics:** Cap: diameter up to 12cm, slimy only when wet, yellow-brown. Tubes: length more than 1cm, olive green; pores very fine. Stem: smooth and firm, length up to 10cm, diameter up to 2cm, light brown. Flesh: yellow, hardly changes colour, has a striking, sour scent, by which only experienced mushroom collectors can identify it.—**Habitat:** A very common mushroom associated with the pine (mycorrhiza), which it follows from the lowlands to the mountains.—**Possibilities of confusion:** Look for the colour of the tubes and note the typical scent.—**Use:** Suillus variegatus is an edible mushroom, which, compared with other tubular fungi, especially Boletaceae, has little value. Only young specimens with firm flesh should be used; suitable for casserole dishes.—**Relatives:** Suillus piperatus, with a peppery taste and a yellow end of the stem, suitable as a seasoning mushroom.

The Cow Boletus

Suillus bovinus

The Cow Boletus grows in close association with Gomphidius roseus as shown in the illustration.—**Characteristics:** Cap: diameter up to 10cm, wavy, irregular, slimy, often coalescing. Tubes: fairly short, cannot be detached from the cap, olive colour. Stem: length up to 6cm, several growing in clusters, tough to elastic, yellow, often red at the base. Flesh: yellow, soft, rather elastic, rubbery. A cap resumes its original shape after having been squashed.—**Habitat:** Mainly in sandy pine forests, but also to be found in Alpine meadows, summer to late autumn.—**Possibilities of confusion:** The strikingly elastic cap makes its distinction from the other slimy tubular fungi easy. May be confused with the much rarer Gyrodon lividus, which however, grows under the alder (Alnus), i.e. mostly in riparian forests, and is also edible.—**Use:** Because of its rubbery texture not highly valued, but like Gomphidius roseus suitable for mixing with other mushrooms.

The Rough-stemmed Boletus

Leccinum scabrum

If you come across a group of birch trees in a wood, or pass even a solitary birch, make a point of carefully examining the soil around the white tree trunks, because it is the favourite habitat of a number of fungi belonging to various genera. You will find almost a dozen of them, in addition to the Fly Agaric, associating almost exclusively with the birch, hardly to be found under any other tree. The Rough-stemmed Boletus is one of them.—**Characteristics:** Cap: average diameter 12cm, hemispherical at first, flat later, moist, in various shades of brown. Tubes: length more than 2cm, distinctly protruding downwards in old specimens, sinuate round the stem. The pores, initially white, acquire a light pink sheen with advancing development. Stem: length up to 18cm, comparatively slender, tough and fibrous later. The whole stem is covered with brown to black scales on a white surface. Flesh: granular, firm in young mushrooms, but soon becoming spongy, particularly in the cap. White, but slowly turning very slightly red, and sometimes green at the base of the stem.—**Habitat:** Common during summer and autumn mostly under birch trees, and therefore also on moors, where it is often very pale. Found throughout Europe and Asia.—**Possibilities of confusion:** Only one Rough-stemmed Boletus is described in old books on fungi. More recent research has revealed the existence of half a dozen similar species. Because all of these are edible any possible confusion is unimportant.—**Use:** Wholesome edible mushroom provided it is young and has not yet turned soft and spongy. Stems of old specimens should be discarded. During preparation for the table the flesh loses its white colour, becoming grey to black without, however, losing its value.—**Relatives:** Leccinum duriusculum, stout like the Red Boletus rather than slender, found under aspens (Populus tremulae), edible. Leccinum griseum, with flesh discolouring a vivid violet-black, always grows under hornbeams or aspens, also edible. As members of the genus Leccinum the Red and the Rough-stemmed Boletus share the characteristics of scaly stems.

The Red Boletus

Leccinum testaceo-scabrum

The members of the genus Leccinum have dry caps, a mostly grey layer of tubes, distinctly sinuate round the stem, which is covered with conspicuous scales. They inhabit extensive regions of the Northern Hemisphere, and as mycorrhiza fungi play an important part in afforestation schemes. They are widely exported by Eastern European countries.—**Characteristics:** Cap: diameter from 5 to 20cm. The skin is orange-coloured and protrudes beyond the margin as a membranous fringe. Tubes: various shades of grey, always sinuate, growing to a length of about 2.5cm. Stem: surface white, dense, covered with black scales, often tapering towards the top. Flesh: white, soon turning light blue or red, has a pleasant scent and taste.—**Habitat:** A common fungus confined to the birch, with which it is mycorrhizal. From lowland plains to the lower Alps it is found throughout Europe and in those parts of Asia and North America which have a similar climate; mainly between June and October.—**Possibilities of confusion:** In the past mushroom experts were unaware of the fact that what they thought were specimens of the Red Boletus do not belong to one and the same species, but are members of different varieties. Because each of these are fit to be eaten, confusing them does not matter.—**Use:** The Red Boletus is a very popular edible fungus, which can also be bought in the market in some continental countries. Unlike many other tubular mushrooms the Red Boletus is usually left alone by the maggots of the Mushroom Fly. Many housewives dislike the fact that its flesh turns black during cooking whereas that of the Edible Boletus remains white; but this discolouration does not indicate that the mushroom is rotten, let alone poisonous.—**Relatives:** Occasionally a variety of Leccinum with a rufous (fox-coloured) cap and flesh that is blue-green around the base of the stem is found under pine trees, often among bilberries; this is Leccinum vulpinum, which is edible. Leccinum aurantiacum with reddish instead of black scales, grows under aspens; it, too, is a prized edible mushroom.

Strobilomyces floccopus

The first encounter with Strobilomyces floccopus causes great surprise. Its appearance is so different from what we expect a tubular fungus to look like that it is readily appreciated that the mycologists have placed it in a family of its own, the Strobilomycetaceae. The unusual dark-brown colour and the cap with its rough scales make the fungus look quite exotic. It is indeed the only representative of its family which is at home in Europe; all the other eight members are confined to the warmer parts of other continents.—**Characteristics:** Cap: the rough, dark brown scales are conspicuous, covering the cap, which has a diameter of 12cm, like roof tiles. A white, woolly veil extends from the margin of the cap to the stem before the cap opens up. Tubes: length up to 3cm, with slighly sinuate and slightly decurrent angular pores. Stem: length up to 15cm, slender, grey to black, with a grey, woolly, frayed ring, the remains of the original veil. Flesh: fairly soft in the cap, more woody in the stem, white, turning first red and then grey to almost black when cut.—**Habitat:** Strobilomyces floccopus is found usually in beech woods, where it favours chalky soil. It is quite common but easily overlooked because of its inconspicuous colour. It is often gregarious and grows in the same location year after year from July to October.—**Possibilities of confusion:** Strobilomyces floccopus can hardly be confused with other tubular fungi. It is, however, possible for a beginner who looks at the fungus casually from above, i.e. does not see the tubular layer, to confuse it with Sarcodon imbricatum (p. 122 bottom), which has a similar scaly but brown cap and spines instead of tubes.—**Use:** Edible, but not very popular.—**Relatives:** A tubular fungus reminiscent of Strobilomyces floccopus with its dark brown colour and drab appearance but with a wholly smooth cap is found quite often in our forests; it could be mistaken for a dark version of the Edible Boletus, but it is Porphyrellus pseudoscaber, also edible but not very tasty. In the U.S. Strobilomyces floccopus is known as the "Old Man of the Woods".

Paxillus involutus

In the past a large number of fungi with caps whose margin is bent to form a distinct rim were classed as members of the genus Paxillus merely on outward appearances. The genus has been reduced to about four species since microscopic characteristics have been taken more into account.—**Characteristics:** Cap: diameter up to 15cm, involuted along the margin, a little grooved, ochre, fine felt-like texture, centre slightly depressed. Gills: easily detached from the cap, colour of yellow wood, decurrent; immediately react to pressure by turning brown. This sensitivity is typical. Stem: length up to 5cm, brown, solid flesh. Flesh: soft, juicy, slightly sour taste, pleasant scent and flavour.—**Habitat:** Common in the most diverse locations such as mixed woodland.—**Possibilities of confusion:** Paxillus filamentosus differs in that its cap is scaly and it grows under alders in Central Europe.—**Use:** In earlier books on fungi Paxillus involutus was listed as edible. More recent books insist that it should be cooked for a prolonged period. Latest research has shown that an occasional meal of Paxillus involutus does no harm, but consumption of the fungus continued over a number of years may lead to a very serious illness and even death.

Paxillus atrotomentosus

Paxillus atrotomentosus is similar to Paxillus involutus.—**Characteristics:** Cap: larger than that of Paxillus involutus, diameter up to 25cm. Gills: yellow. Stem: unlike that of involutus, short, diameter up to 5cm, covered with a brown, velvety felt. Flesh: Unusually high water content. When the fungus is squeezed, much fluid oozes out.—**Habitat:** Very common on tree stumps but also on woodland soil.—**Use:** Paxillus atrotomentosus, although not poisonous, is not particularly palatable, and therefore rarely collected for the table.—**Relatives:** Paxillus panuoides, of eccentric, shell-like growth on wood is not recommended.

Gomphidius glutinosus

In woods, fungi are often found which are covered with a thick layer of slime either on the cap or on the entire surface. They belong to the genera Phlegmacium, Myxacium, and Suillus. Because of their sliminess these fungi are often shunned by collectors, although some of them, such as Gomphidius glutinosus, are first-class eaters.—**Characteristics:** Cap: cool and soft to the touch, grows to a diameter of 10cm, is covered with a layer of slime which in young specimens extends to the stem and is easily peeled off. Older caps often have black patches. Gills: at first white, but turning black as they become covered with spores, very decurrent and bifurcate towards the margin of the cap. Stem: length up to 10cm, becoming thicker towards the top, slimy, with remnants of a ring; its yellow base is very conspicuous. Flesh: very soft and at the base of the stem a brilliant lemon colour, which is an infallible distinguishing mark.—**Habitat:** Common in every pine forest whether naturally grown or planted, because it lives in symbiosis with the pine. Summer to autumn.—**Possibilities of confusion:** No other fungus has flesh in the stem that is as brilliant yellow and a cap with as shiny a surface as Gomphidius glutinosus.—**Use:** Unfortunately this good edible mushroom is not as universally valued as it deserves to be, because it tends to stick to other mushrooms in the basket; the slimy skin should therefore be peeled off the cap as soon as the mushroom is gathered. It is altogether advisable to carry and store it apart from other items of the crop, because older specimens secrete an abundance of spores, which contaminate the other mushrooms.—**Relatives:** 18 species of Gomphidius are found throughout the world; Several more of these grow in Britain, including Gomphidius roseus and Gomphidius maculatus, which associates with the larch; both are suitable for the table.

Camarophyllus pratensis

Characteristics: Cap: shaped like a spinning top, diameter up to 10cm, flesh thick in the centre, thin along the margin, orange-yellow, with a small boss (hump) in the centre of the cap. Gills: thick and waxy, distinctly decurrent, colour similar to that of the cap. Stem: length up to 10 cm, narrowing towards the ground, progressively broader towards the cap, paler than the cap. Flesh: no noticeable scent and taste; red-yellow.—**Habitat:** Camarophyllus pratensis appears comparatively late in the year, not before Spetember, lasting to November. It hardly enters the interior of forests, but is found on pastures and meadows. In the Alps it can be seen at altitudes exceeding 2,000m; it grows throughout Europe and Asia.—**Use:** A valuable edible fungus, not usually attacked by insect larvae.

Hygrophorus marzuolus

If the papers published an item on April Fools' Day about mushroom collectors that had been seen returning from the woods loaded with baskets full of mushrooms it would probably be treated as a joke. The report might nevertheless be true, because Hygrophorus marzuolus is one of the fungi that have chosen such an unusual time to make their appearance; it is much talked about, but found by few.—**Characteristics:** Cap: diameter up to 12cm, at first white, then grey to black; dry, often of very irregular shape, thick flesh. Gills: of most conspicuous thickness, waxy, distant (widely spaced), slightly decurrent. Stem: hardly longer than 6cm, comparatively thick, white. Flesh: white, grey under the skin of the cap, of mild flavour without distinct scent.—**Habitat:** Often found in small groups on the Continent (but not in Britain), especially in mixed forests in the mountains. Depending on weather conditions from December to May, always found as soon as the snow melts. Inconspicuous because it often fails to break through, remaining more or less covered by a layer of leaves.—**Possibilities of confusion:** Hygrophorus camarophyllus is similar, but grows only in autumn and has a distinctly fibrous cap; it, too, is edible.—**Use:** Substantial, choice edible fungus in spring, a season poor in mushrooms.

Hygrocybe punicea

Characteristics: Cap: at a diameter of 12cm the largest and, on account of its colour, varying from yellow to blood-red, one of the most beautiful fungi known; moist, waxy and very fragile. Gills: thick and waxy, distant, at first yellow, later orange-coloured. Stem: strong, length up to 10cm, diameter more than 2cm, hollow, top yellow, base white. Flesh: white in the centre, yellow towards the margin, watery, not very substantial, without notable scent or taste.—**Habitat:** Hygrocybe punicea is a typical autumn fungus which does not appear before September and October. It is common in Britain, and found in meadows and pastures at the edge of forests, in the grass of forest glades, especially in higher altitudes.—**Possibilities of confusion:** Large, adult specimens can hardly be confused, because no other Hygrocybe grows so tall; smaller specimens may be confused with other members of the genus.—**Use:** Edible like all other mild-tasting Hygrocybes.

Hygrocybe conica

The genus Hygrocybe strikes the eye with its gorgeous colours, mostly brilliant red and yellow. Its members have a waxy consistency. Hygrocybe conica is a typical representative.—**Characteristics:** Cap: conical, pointed, never flat, vivid red and yellow colours, margin often torn, distinctly black when old or damaged. Gills: adnexed, white at first, yellow later, finally black, edges serrated. Stem: hollow, fragile, same colour as that of the cap, often twisted. Flesh: mild flavour, without distinct scent; white, yellow towards the margin.—**Habitat:** Common in forest clearings and damp meadows and pastures, summer to autumn, in the whole of Europe and Asia.—**Possibilities of confusion:** About 40 different species of Hygrocybe are found in Europe, some of which resemble each other very closely. Although none are poisonous, a few are bitter and therefore unpalatable.—**Use:** Like all the species of Hygrocybe that are not bitter, Hygrocybe conica is an edible mushroom, but because it is unsubstantial is little collected.

Clitocybe nebularis

Clitocybe nebularis is a member of a large genus, represented in Britain alone by about 70 species. Clytocybes never have a ring round the stem or other remnants of a veil, and their stems are often more or less fibrous. In many old specimens the cap is funnel-shaped. The gills are thin, either adnate (grown to the stem) or decurrent.—**Characteristics:** Cap: at first hemispherical, later spread out but unlike the typical Clitocybe form, never funnel-shaped; umbilicate; involuted margin, attains a maximum diameter of 15cm, various shades of grey and often covered with a bloom in which scattered plant detritus becomes embedded. Gills: white, turning ochre later, slightly decurrent and very easily detached from the flesh of the cap, which is a very prominent characteristic. Stem: strong; longitudinal fibrils, length up to 12cm, diameter 3cm, white to grey. Flesh: thick in the cap, white, does not become discoloured, has a penetrating smell, which is not to everyone's liking.—**Habitat:** Clitocybe nebularis is very common and grows in long rows or circles, the so-called fairy rings. It is found only during the autumn months, but then in profusion. It grows in coniferous forests, but also in the litter of beech leaves throughout Europe and Asia.—**Possibilities of confusion:** Its size, the detachable gills, the penetrating smell and the late appearance make it easy to distinguish from other members of the genus. Clitocybe alexandri is similar, but its gills are darker, and the base of the stem is woolly. This double is rare and as an edible mushroom hardly significant.—**Use:** Clitocybe nebularis is an edible mushroom and in the past was sold in markets on the continent. Because it is very fleshy and found in abundance it enjoyed a certain popularity. It has, however, been established that it does not agree with everybody, and different members of a family may react differently to it. This had produced very divergent judgements of its value. Before it is eaten for the first time it should therefore be tested in small quantities for its digestibility.

Fammulina velutipes

The question whether mushrooms can be grown in one's own garden or allotment is frequently asked. For most types of mushroom this possibility must be ruled out. Apart from cultivated button mushrooms only a few kinds, usually growing on wood, such as the Oyster Mushroom, Pleurotus ostreatus, lend themselves to this. Fammulina velutipes is another.—**Characteristics:** Cap: reaches a diameter of 5-7cm, hemispherical at first, flattening later, wavy and irregular when old, brown, yellow towards the margin, in damp weather distinctly slimy; glutinous and sticky. Gills: yellow, later darkening slightly, easily removed, of varying length. Stem: length up to 8cm, often flattened, becoming hollow later. It is covered with a dark brown, almost black, velvety felt right up to the cap. Flesh: rather light-coloured, without typical scent or flavour.—**Habitat:** Always on wood, trees, or tree stumps, rarely on fallen branches. It clearly prefers deciduous trees, mainly willows (species of Salix). Unlike other agarics, it is found from December to March, and in abundance when wintery weather gives way to thaw and mild westerly airstreams. Even during hard frost it keeps fresh under a cover of snow and survives when thawed out. Its growth in clusters is typical.—**Possibilities of confusion:** There is no risk of confusing it with other fungi that grow in clusters; among these is one, the Sulphur Tuft, Hypholoma fasciculare (p. 96 bottom), which must be carefully distinguished from Fammulina velutipes, because it is not only bitter, but has only very recently been found to be poisonous. Its yellow-green gills are an unmistakable distinguishing mark.—**Use:** Because there are hardly any fresh mushrooms found about Christmas time, Fammulina velutipes is a popular and highly prized edible fungus. It can also be cultivated.

Collybia dryophila

The genus Collybia includes small to medium-sized species, usually with a gristly, tough stem; they are almost totally ignored by collectors of edible fungi. Collybia dryophila is one of them, and deserves much more attention.—**Characteristics:** Cap: diameter up to 7cm, flat, wavy, yellow, flesh-coloured towards the margin, which in watery specimens is fluted. Gills: white, very crowded (densely spaced) and adnate (attached to the stem). Stem: tubular, with longitudinal grooves, length up to 8cm, red-brown, at the base white, felt-like. Flesh: not very substantial, very thin, has a mild taste and pleasant scent.—**Habitat:** Very common at all altitudes in summer and autumn, growing in large clusters of many individuals.—**Possibilities of confusion:** Collybia peronata looks similar, but is distinguished by the burning, sharp taste of a sample, which is harmless.—**Use:** Although Collybia dryophila is not an edible mushroom of the first rank, and because of its thin flesh not very substantial, it is not altogether worthless as an addition to a dish of mushrooms.

Collybia butyracea

The adjective butyracea does not mean that the mushroom has the colour or, when cooked, the softness of butter; it merely indicates that the skin of the cap has a greasy sheen. The base of the stem is club-shaped, resembling an inverted carrot in shape.—**Characteristics:** Cap: average diameter 8cm, little flesh, smooth, with a broad central boss, greasy sheen, red to chestnut-coloured. Gills: crowded, adnate, slightly notched edges, white, sometimes pale red. Stem: never consists of solid flesh, always hollow, tapering from the base upwards, longitudinal fibrils, gristly, rigid, a little paler in colour than the cap. Flesh: white, watery, not very substantial.—**Habitat:** Very common particularly in mossy pine woods from summer to late autumn.—**Use:** Edible apart from the stem, not highly valued. It also has a grey variety.

Trichomolopsis rutilans

This gorgeous fungus is a typical example in which the appearance of the young specimens differs so widely from that of the older ones that they are thought to be different species.—**Characteristics:** Cap: average diameter 15cm, at first hemispherical, later flat. The basic colour is yellow, but initially the cap is so densely covered with a purple felt that it appears evenly red. When it opens up the scales separate and the basic yellow colour appears. When all the scales have been washed away by the rain, all that remains is the yellow of the cap. Gills: yellow with ragged edges. Stem: grows to a length of 12cm and a diameter of up to 2cm, red scales on yellow background. Flesh: yellow, juicy, mild.—**Habitat:** On conifer stumps from summer to autumn, even in dry weather.—**Possibilities of confusion:** Typical specimens are easily identified by their colour, their yellow flesh and gills, and the fact that they grow on wood.—**Use:** In spite of its gorgeous appearance it does not rank high as an edible fungus. At best to be used with other mushrooms.

Tricholoma flavovirens

In regions with extensive deciduous forests on chalky soil Tricholoma flavovirens is almost unknown, but found in abundance in large pine forests on sandy soil.—**Characteristics:** Cap: diameter about 10cm, young specimens with involute margin, with a slight boss, ingrown fibrils; sulphur-yellow, boss usually brown. Gills: yellow, deeply sinuate. Stem: length up to 6cm, diameter more than 1cm, often deeply grown into the sand. Flesh: rather fleshy round the boss, white, yellow under the skin of the cap, has a vaguely mealy smell and mild taste.—**Habitat:** Widely distributed mainly on sandy soil in pine forests, towards autumn.—**Possibilities of confusion:** Tricholoma sulphureum is very similar, but easily distinguished by its most offensive smell of burned rubber, town gas, or carbide; it is poisonous.—**Use:** Tricholoma flavovirens is a very popular edible mushroom; grains of sand must be carefully removed.

St. George's Mushroom

Calocybe gambosa

Classed by earlier mycologists as a true Tricholoma, Tricholoma Georgii, the St. George's Mushroom. Microscopic examination has, however, shown that only its outward appearance suggests this classification; its microstructure is not that of a Tricholoma. The genus Calocybe has therefore been established for it and for similar fungi.—**Characteristics:** Cap: diameter about 10cm, occasionally larger, at first globular with involute margin, later flat, often irregular, white to yellow, sometimes ochre. Gills: white, very crowded, sinuate, sometimes decurrent with a small tooth. Stem: strong, base sometimes club-shaped, white to yellow, firm flesh, length up to 8cm. Flesh: smells strikingly of meal, which is an important mark of recognition.—**Habitat:** In meadows, often in fairy rings, April (23rd, St. George's Day) to May.—**Possibilities of confusion:** With Inocybe patouillardii (p. 86 top), which appears at the same time and is poisonous; the distinguishing features of Inocybe patouillardii are the pointed conical cap and flesh that turns deep red.—**Use:** A valued edible mushroom.

Tricholoma pardinum

Every fungus collector in Europe should know Tricholoma pardinum, because it is a dangerous poisonous species; for instance, it causes most cases of mushroom poisoning notified in Switzerland.—**Characteristics:** Cap: hemispherical, later almost funnel-shaped, grey to brown, coarse scales, diameter up to 15cm. Gills: white, sinuate, sometimes exuding a fluid. Stem: rather strong, length up to 8cm and diameter more than 3cm, firm flesh. Flesh: white, but turning grey under the skin of the cap and red in the base.—**Habitat:** Mainly in beech woods on chalky soil, common in some regions but not everywhere.—**Possibilities of confusion:** Similar, more slender, fine-scaled species of Tricholoma are collected as edible fungi. In cases of doubt they should be shown to experts, because distinction is not easy.—**Use:** Should on no account be used as an edible fungus, because it produces violent gastro-intestinal upsets; however, lethal cases are not known.

Laccaria amethystina

This strikingly handsome mushroom has baffled many fungus collectors as well as taxonomists. Many that gather mushrooms for the table cannot accept that such a deep-violet fungus could possibly be edible. Taxonomists were for a long time at a loss how to classify a mushroom that has spores like Russula, thick gills like Hygrophorus, and a deeply umbilicate cap like a typical Clitocybe. This is how the genus Laccaria came to be set up for Laccaria amethystina and its nearest relatives.—**Characteristics:** Cap: at first very convex, flattening later, umbilicate in the centre, with fine scales, deep violet when young, but later bleaching to almost white. Gills: distant, distinctly decurrent, of uneven length. Whereas the cap becomes pale, the gills remain an attractive violet for a long time. Stem: length up to 8cm, diameter up to 1cm, often slightly bent, longitudinal fibrils. Same colour as that of the cap and gills when young, but becoming pale later and perhaps turning brown in the lower part. Flesh: amethyst-violet like the whole mushroom, without distinct scent and taste.—**Habitat:** Very common in the most varied types of woodland, but clearly preferring deciduous forests; found from summer to late autumn.—**Possibilities of confusion:** Typical specimens are of a colour that makes it almost impossible to confuse them with any other fungus. Somewhat dried and therefore bleached fruit bodies resemble Mycena pura which is poisonous because it contains muscarine. In cases of doubt examine the smell of the fungi, because the poisonous Mycena pura smells distinctly of radishes, whereas Laccaria amethystina gives off no scent except when old, when it may smell sweet.—**Use:** In spite of its colour edible, but not particularly valuable.—**Relatives:** A mushroom, Laccaria laccata, which is almost indistinguishable from Laccaria amethystina, is found even more often, but it is reddish and extremely variable in colour and size. Its table value is the same as that of Laccaria amethystina, i.e. it is also edible, but of mediocre quality.

The Honey Agaric

Armillariella mellea

Many fungi growing in symbiosis with trees promote the development of their partners, i.e. are useful to silviculture. This cannot be said of the Honey Agaric because it is a parasite which attacks living trees, feeding on its host and ultimately killing it. The forester looks at it through jaundiced eyes, unlike the fungus collector, whom it often provides with abundant crops gathered without effort; it can, as it were, be picked up in passing..—**Characteristics:** Cap: average diameter 10cm, brown, sometimes honey-coloured, hence its name. At first covered with dark, bushy, hairy scales, but becoming bald later. Gills: white, turning yellow-red, with a white dust of spores, fairly distant, slightly decurrent. Stem: length about 15cm, often curved, yellow to brown. The remnants of the veil, which in the young specimen extends from the stem to the margin of the cap and at first completely covers the gills, form a ring round the upper part of the stem. Flesh: fibrous in the stem and tough. The flavour, typical of the species, is striking. If a piece of flesh is chewed the initially mild taste soon turns into a very unpleasant, constrictive, abrasive one. This is a characteristic of the Honey Agaric that enables you to identify it with your eyes closed.—**Habitat:** Very abundant during autumn, growing in clusters of many caps on tree stumps and on roots hidden in the soil, often ascending high up into the trees, which die as a result. It is found throughout Europe and has been distributed worldwide by man.—**Possibilities of confusion:** Once you remember the peculiar, typical flavour, you will always readily recognize the Honey Agaric, although its shape and colour varies widely. Beginners find it difficult to distinguish it from other similar agarics growing on trees, for instance from Kuehneromyces mutabilis (p. 94) and from the Sulphur Tuft, Hypholoma fasciculare (p. 96 bottom), a poisonous fungus.—**Use:** The tough stem is useless. The bitter taste disappears during cooking. Because the Honey Agaric does not agree with everybody, it should be sampled before being eaten as part of a meal.

The Fairy-ring Champignon

Marasmius oreades

Fungi belonging to the genus Marasmius have the property of losing a large amount of water through evaporation in dry weather, shrinking in the process; after rain they replenish it and revive. It requires some imagination to associate the scent of Marasmius oreades with that of Pinks, although this is widely claimed.—**Characteristics:** Cap: average diameter 5cm, light or dark yellow-brown depending on its moisture content; it has a slight boss. Gills: distant, white, comparatively thick. Stem: length up to 7cm, diameter less than 1cm, colour like that of the cap but lighter, very fine, felt-like texture, rather tough. Flesh: white, with an aromatic scent, reminding people of flowers.—**Habitat:** Not in woods, but in the turf of meadows, pastures, lawns and by the wayside. The fungus grows and lives on dying grass roots. It is found in the same locations year after year in several successive batches, always after rain.—**Possibilities of confusion:** Some similarity exists with several species of Inocybe which, however, produce brown instead of white spore dust and mostly have an unpleasant smell of sperm; almost all are poisonous. —**Use:** Marasmius oreades is a much neglected edible fungus particularly suitable for soups. Although the individual mushroom yields little flesh, it makes up for this by its presence in astonishing quantities in rows or rings.—**Relatives:** In Europe the genus Marasmius is represented by about 30, often tiny, species. Of interest as a choice seasoning mushroom is Marasmius scorodonius, smelling strongly of garlic; it is a small mushroom with a cap of no more than 2.5cm diameter and a bare, horny, red-brown stem, which is white on the upper part. Dried, it is suitable as a seasoning mushroom. It should not be confused with Microomphale perforans, which looks like a small Marasmius scordononius and always grows from a single pine needle; it does not smell of garlic, but is very offensive.

Wood Blewits

Lepista nuda

Snowdrops are the heralds of spring, asters of autumn. Among the fungi, too, there are kinds whose appearance is always associated with a certain season. Lepista nuda is one of them, because to the mushroom lover it announces that the mushroom year is drawing to its close. It was formerly included in the genus Tricholoma, but it differs from it in that its spores are pink; the genus Lepista was therefore established for it.—**Characteristics:** Cap: average diameter 12cm, attractive, convex shape with long involute margin, at first violet but soon turning pale and brown. Gills: still a beautiful violet even after the cap has lost its colour, fairly crowded, of varying length, and easily detached from the flesh of the cap. Stem: length about 10cm, diameter 2cm, colour similar to that of the cap; fibrous. The subterranean mycelium, too, is violet. Flesh: violet when young, turning pale in the old fungus, watery in wet weather, has a pleasant scent.—**Habitat:** In various kinds of wood, usually in fairy rings or long rows from September to late autumn, solitary specimens as early as May; but they can be found frozen stiff as late as November and December.—**Possibilities of confusion:** Must not be confused with other violet agarics of similar appearance. Beginners should remember that possibly confused Cortinarii can be identified by their hairy veils extending from the margin of the cap to the stem, and the distinctly rust-brown colour of the spore dust. Confusion with Cortinarius traganus, although betrayed by its offensive smell, is particularly easy.—**Use:** Lepista nuda is a highly esteemed edible fungus, which is also popular when pickled. Frozen specimens should be rejected. Some connoisseurs are inordinately fond of it; others dislike it because of its sweet taste.—**Relatives:** Lepista personata, with a brown cap and deep-violet stem, edible, and Lepista glaucocana, paler in all parts, hardly recommended. Lepista saera is similar in size and shape with a pale tan or greyish cap and again with pinkish spores. It occurs in pastures and is also good to eat.

Rhodophyllus sinuatus

Rhodophyllus sinuatus is a very large, handsome fungus, of which no more than a few specimens promise a substantial meal. It looks appetizing and its mealy scent, too, creates confidence in the inexperienced collector. It is, however, highly poisonous and produces alarming symptoms soon after having been eaten.—**Characteristics:** Cap: diameter up to 18cm, convex, with a slight boss, flesh very thick in the centre; the margin is dinstinctly involute; ivory-coloured in varying hues, with a silky sheen. The skin can be peeled off the cap. Gills: pink after a long initial period during which they are yellow with a roseate sheen. Rather broad in proportion with the size of the fungus, slightly distant and sinuate. Stem: strong, base club-shaped, length up to 12cm, diameter up to 4cm, becoming hollow when old. Flesh: white, distinctly mealy scent, with a peculiar tang. Sampling it, which is harmless if the small chewed quantity is spat out, gives no indication that it is poisonous, because the flesh has a pleasant taste.—**Habitat:** In some areas Rhodophyllus sinuatus is almost unknown or found in only a few places kept strictly secret by fungus lovers. Elsewhere it may be more common, especially in fertile soil under beech and oak trees, and in warmer climates.—**Possibilities of confusion:** With, for instance, the Horse Mushroom, Agaricus arvensis, (p. 84. top), Clitocybe geotropa and especially Clitocybe nebularis, which, however, usually appears when Rhodophyllus sinuatus is already becoming rare.—**Use:** Comparatively poisonous and in countries such as Switzerland and France where it is common, the cause of a considerable proportion of clinically treated cases of mushroom poisoning. In addition to severe gastro-intestinal upsets, headache, and abdominal pain persisting for days it leads to dizziness and feebleness, but is hardly ever lethal.

Pluteus atricapillus

When it has not been raining for weeks and the dry soil is showing cracks, the prospect of finding fungi is poor, the woodlands are virtually devoid of them. But those who are collectors of some experience know even in times of drought where to look for them to bring home enough to make a meal. Even in dry woodland soil, tree stumps retain a relatively large amount of moisture, enabling some fungi to grow. In Britain Oudemansiella platyphylla, Hypholoma capnoides and Kuehneromyces mutabilis (and on the continent Pluteus atricapillus) are found on tree stumps and decaying wood.—**Characteristics:** Cap: diameter up to 15cm, slightly convex with a small boss after opening up, all shades from light to black brown, with fibrils radiating from the centre. Gills: at first almost white, gradually turning reddish, finally intensely flesh-coloured; they are convex, sinuate and adnexed; this is an important feature of all members of the genus Pluteus. Stem: length up to 12cm, diameter more than 1cm, slender, covered with brown fibrils on a white background; solid flesh. Flesh: rather thin, a little more substantial in the boss of the cap, without distinct scent or flavour.—**Habitat:** Pluteus atricapillus grows only on wood, i.e. it is found on rotting tree stumps and less often on fallen branches throughout the warm season. It prefers stumps of deciduous trees, and thrives even when other fungi have ceased to grow because of drought. Has also been found in cellars on stored wood.—**Possibilities of confusion:** In Britain about three dozen members of the genus Pluteus are found; almost all grow on wood and usually are small and without practical value. Pluteus atromarginatus is very similar, but differs from Pluteus atricapillus in that its gills are black-edged; it, too, is edible and grows on stumps of conifers.—**Use:** Pluteus atricapillus is edible, but rarely collected and not highly prized because it yields little flesh. No poisonous fungus of the genus Pluteus is known.

The Fly Agaric

Amanita muscaria

Pictures of the Fly Agaric are found not only in the scientific literature but also in children's books. It appears with its red cap and white dots as a Good Luck symbol on New Year Cards; in fact, the Fly Agaric is one of the best-known of all fungi. This does not prevent it from being confused with edible mushrooms, and utterly mistaken views about its suitability for the table are advanced—in fact, it is dangerous and poisonous.—**Characteristics:** Cap: the young specimen is enclosed in a white veil, which tears when the cap opens up, revealing the brilliant red cap (occasionally slightly yellow), which is covered with white dots, the remnants of the veil, on the red background. Reaches a diameter of up to 20cm. Under dripping trees the remnants of the veil may be washed off altogether, leaving the cap a uniform red. The margin is slightly fluted. Gills: white, free, i.e. they do not reach the stem. Stem: length up to 20cm, diameter up to 2cm. The base is encircled by several superimposed warty rings, remnants of the universal veil. A white cuff surrounds the upper part of the stem.—**Habitat:** The Fly Agaric is common in coniferous woods and particularly under birch trees during summer and autumn. It is found in lowland and mountain regions throughout the Northern Hemisphere.—**Possibilities of confusion:** Young Fly Agarics which do not yet show any red are easily confused with the Common Puff Ball, Lycoperdon perlatum (p. 130). In cases of doubt, cut the fungus in half. A red-yellow line, which is absent from the Puff Ball, appears in the Fly Agaric.—**Use:** The Fly Agaric is highly poisonous. Admittedly the poison content varies, and in the East the Fly Agaric is eaten for intoxicating purposes. But the claim that Amanita muscaria is completely harmless after the removal of the red skin of the cap is untenable. Poisoning leads to fits of delirium, and disturbances of vision and speech, but it is rarely lethal. Milk poisoned with the Fly Agaric was in the past used to kill flies, which explains the name of the fungus.—**Relatives:** The Fly Agaric is very mutable. A related form, Amanita muscaria, with a brown instead of a red cap, is equally poisonous and found mainly in coniferous forests in mountains on the continent.

Destroying Angel, Fool's Mushroom

Amanita verna = Amanita virosa (the Destroying Angel)

Agarics that are pure white all over, cap, gills, and stem, should always be examined with the greatest possible care, because the deadly poisonous Fool's Mushroom is also pure white.—**Characteristics:** Cap: often conical, diameter up to 10cm, sticky, with a dry sheen, only in the centre light ivory-coloured. Gills: pure white with flaky edges. Stem: length up to 15cm, growing from a membranous sheath, with a white ring round the upper part, which in old specimens may disappear. Flesh: white, has an unpleasant smell. It should never be sampled.—**Habitat:** In deciduous and coniferous woods as early as July, to autumn.—**Possibilities of confusion:** Fatal cases of poisoning are caused when collectors confuse the Horse Mushroom and the Field Mushroom (p. 84) with the Fool's Mushroom; but in the first two the gills are white only in quite young specimens and soon become dark. Nor do the Horse Mushroom and the Field Mushroom have a sheath at the base of the stem. Symptoms of poisoning do not appear until 11-12 hours after consumption and seriously endanger the patient's life.

The Golden Agaric

Amanita caesarea

True fungus lovers do not confine their interest to the musroom season during summer and autumn, but pursue it throughout the year including holidays; if these are spent in the Sunny South, they will find species they will never come across at home. One of these is the Golden Agaric.—**Characteristics:** Found for the first time, it suggests a Fly Agaric, but clear distinctions are soon noted. Cap: brilliant red to orange-coloured, without white spots. Gills: a gorgeous golden yellow; of varying length. Stem: golden yellow, growing from a white sheath, which at first enwraps the whole mushroom, making it look like a white egg when young. Stem surrounded by a yellow cuff. Flesh: yellowish.—**Habitat:** In warm regions, of Southern Europe.—**Possibilities of confusion:** Dangerous where Golden Agaric and Fly Agaric are found together; at the egg stage it may even be confused with the Fool's Mushroom.—**Use:** A highly prized, first-class delicacy. North of the Alps this mushroom is so rare that in the interest of conservation it should not be collected at all.

The Panther Cap or False Blusher

Amanita pantherina

The Panther Cap has caused many cases of poisoning because in old fungus books it was often confused with Amanita spissa, which is edible. Illustrations of the latter bore the legend "Panther Cap".—**Characteristics:** Cap: diameter up to 10cm, hemispherical, later flat, the skin can be peeled off, with small white flakes in concentric rings, margin fluted, light brown. Gills: white, crowded. Stem: length up to 12cm, white, hollow, growing from a non-lobed bulb (volva) as from a flowerpot, with a white, non-grooved, perfectly smooth cuff. Flesh: white, smelling of radishes.—**Habitat:** Common in pine woods on sandy soil. Occasional in Britain. Summer and autumn.—**Possibilities of confusion:** With the edible Amanita spissa, whose ring is distinctly fluted (most important mark of recognition).—**Use:** The Panther Cap is dangerously poisonous and owing to confusion with Amanita spissa causes many cases of poisoning, which, however, are not lethal.

Amanita spissa

Characteristics: Cap: diameter up to 12cm, at first hemispherical, slightly convex after opening up, grey brown, covered with small white flakes. The margin is hardly fluted. Gills: white, with flaky edges, sinuate. Stem: often stout, thick-set, with a ring that is very fine but distinctly fluted like a finely pleated ladies' skirt; sometimes you may not be able to see this without the aid of a magnifying glass. The base is bulbous. Flesh: white, discolours light brown.—**Habitat:** Not commonly found in deciduous and coniferous woods in Britain.—**Possibilities of confusion:** With the Panther Cap (often). Old specimens, which have lost the ring, or where it is so wilted that it is no longer possible to see whether or not it is fluted, should be rejected.—**Use:** Amanita spissa is an edible fungus collected by experts. Because it widely varies in appearance, is only moderately tasty and easily confused, beginners are warned to leave it alone.

The Death Cap

Amanita phalloides

The Death Cap is one of the most dangerous poisonous fungi in our woods. Like the Destroying Angel, Amanita virosa, it has caused the deaths of many people, of entire families, and on one occasion of 31 children in a holiday camp. Every fungus collector should know the Death Cap. It is so dangerous because the symptoms of poisoning do not appear until the day after the meal, when the toxic substance has already entered the bloodstream and laxatives and emetics are no longer effective. The improvement in the patient's condition on the third day is deceptive.—**Characteristics:** Cap: the young fungus is entirely covered by the universal veil and looks like a white egg. Only when the veil bursts open will the cap be revealed. It is bell-shaped at first, and when opening up reaches a diameter of 15cm. It is somewhat sticky, with a hint of green, with dark, ingrown radial fibrils. It rarely has remnants of the universal veil. Gills: white, later with a greenish sheen, adnexed or free. Stem: length up to 15cm, white, with a drooping cuff, below which it shows a very delicate snakeskin pattern with fine scales. Flesh: white, with a green tinge below the skin of the cap, does not discolour in air. The base of the stem is surrounded by a lobed, membranous sheath.—**Habitat:** Found in oak and beech woods from July to October.—**Possibilities of confusion:** Confusion with similar but edible fungi creates the danger of lethal mushroom poisoning. If you find an agaric with a green cap, take it out of the ground entirely; if you cut it off, you will leave a vital clue to its identification, the membranous sheath, in the soil. Green, edible agarics with which it is confused have neither a ring nor a sheath round the base of the stem. Tricholoma flavovirens (p. 52 bottom), with which it has also been confused, is a much collected edible fungus with distinctive yellow gills; it too, has neither a ring nor a sheath.

Suspected Death Cap poisoning demands immediate clinical treatment, if possible in a special toxicological department, established in almost every hospital.

The Blusher

Amanita rubescens

The number of edible fungus species collected in the past was small. Hardly anything was gathered except the Edible Boletus, the Red Boletus, the Rough-stemmed Boletus, and the Chantarelle. Agarics were always suspect and eating a member of the genus Amanita would have been considered foolhardy in the extreme. Today this would be exaggerated caution, but it was well justified in the past because of the lack of reliable information. The white colour of the spores distinguishes the Blusher from the dark-spored mushrooms (Agaricus).—**Characteristics:** Cap: average diameter 12cm, at first hemispherical, after opening up slightly convex, reddish brown but often very pale, with many white warts, which are sometimes washed off by rain. Gills: at first white, later reddish, crowded. Stem: length up to 15cm, usually very strong, colour similar to that of the cap, but always more pale. It is surrounded by a large, white, fluted cuff. At the base the stem is bulbous, often with a few rather indistinct scaly rings. Flesh: white, but after injury always turns a more or less bright red; this is an important distinguishing sign. Without marked scent and flavour.—**Habitat:** Very comnon in deciduous and coniferous woods, and can be collected in large quantities during the summer and autumn months.—**Possibilities of confusion:** The Blusher is very variable. Side by side with very stout specimens grow some that are so slender that the uninitiated will find it difficult to believe that all these widely varying forms are members of one and the same species. Such poor specimens often have a yellow cuff, but they, too, turn red, exhibiting therefore the decisive characteristics of the Blusher. None of the highly poisonous Amanitae has flesh that turns red.—**Use:** The Blusher is a delectable, substantial edible fungus, which should, however, not be eaten raw. It is unfortunately very often attacked by the maggots of the Mushroom Fly.—**Relatives:** In some fungus books another "False Blusher", Amanita pseudorubescens, is mentioned, which was said to be poisonous. It is described as dark, grey violet, but is perhaps only a form growing in extremely dry habitats. Such unconfirmed suspicions should not be allowed to diminish the esteem in which the Blusher is held.

The Grisette

Amanita (Amanitopsis) vaginata

In the past, the Latin name of the Grisette was Amanitopsis, i.e. looking like an Amanita; but it differs from it in that it has no ring round the stem. When it was realized that in the earliest youth rudiments of a ring are present, the Grisette was re-classified as a member of the genus Amanita.—**Characteristics:** Especially the colour of the cap is so variable that the Grisette can be said to belong to half a dozen different species or subspecies. Cap: average diameter 10cm, with a small but distinct boss. The radial grooves in the margin are a prominent feature. Young mushrooms are completely enclosed by a white universal veil like an egg by its shell. The colour of the cap varies widely from white to grey, yellowish, red-brown, dark brown. Gills: white, free. Stem: length up to 12cm, slender, hollow, very fragile, colour similar to that of the cap or banded pattern, always without a ring, growing from a membranous, often torn sheath at the base. Flesh: soft and white, without any distinctive features.—**Habitat:** Worldwide, in its various forms in all types of wood from the lowlands to the mountains.—**Possibilities of confusion:** The greatest possible care is essential when pure white Grisettes are being collected, to prevent Fool's Mushrooms or Destroying Angels from joining their company by mistake. The poisonous fungi differ in that they have rings, and the margins of their caps are not grooved. Very young Grisettes, which are still egg-shaped, must also be carefully examined. The membranous veil should be scraped off and the future margin of the cap inspected for distinct grooves, which are evident even in very young specimens.—**Use:** Grisettes are edible, but because their flesh is very thin are not very substantial. The various colours of the cap have no bearing on their edibility. Not wholesome when raw.—**Relatives:** Occasionally a very large Grisette is found, with thick, grey remnants of the veil on the cap and a multiple sheath surrounding the stem. This is the Giant Grisette, Amanita inaurata, also a useful edible fungus.

The Parasol Mushroom

Macrolepiota procera

The Parasol Mushroom has a most appropriate name, because hardly another agaric reaches a diameter of 30cm. Parasol aptly describes its shape. In the past it belonged to the genus Lepiota. Very large spores, the anatomical structure of the skin of the cap, the movable ring, and the coarse scales on the cap decided the mycologists to establish the genus Macrolepiota for it and its relatives.—**Characteristics:** Cap: at first oval, giving the fungus the shape of a drumstick before it opens up. After the cap has spread and flattened, its diameter may reach 30cm. Whereas the centre of the cap remains smooth and brown, the marginal portions of its skin disintegrate into coarse scales, revealing the light-coloured flesh in between. Gills: the broad, white lamellae are free. Stem: club-shaped, drumstick-like base, length up to 30cm, slender, brown snakeskin pattern. A very distinctive ring is an important identification mark; with care it can be moved along the stem. Flesh: remains white when injured, gives off a pleasant scent and has a nutty flavour.—**Habitat:** In clearings, on the edge of woods and in meadows nearby rather than in dense woods, during summer and autumn. Although the fungus is not associated with any particular tree, it will be found in the same places year after year.—**Possibilities of confusion:** In its typical aspect hardly to be confused with another large Parasol Mushroom. Very young specimens, hardly 5cm tall, look so odd that a beginner will certainly not expect them to develop into ordinary Parasol Mushrooms.—**Use:** The Parasol Mushroom is a highly praised edible fungus which, like its relatives sold in the markets of Africa, Asia, and Europe, is recommended especially when sliced. The woody stems, however, cannot be used.—**Relatives:** A smaller "Parasol Mushroom" found assumes, to the collector's surprise, an intense saffron colour when cut. This is Marcrolepiota rhacodes, grows in woods, is smaller than the Parasol Mushroom, and can be used in the same way.

Agaricus silvaticus

A considerable number of cases of mushroom poisoning is due to the confusion of Agaricus species with the Fool's Mushroom, the Destroying Angel, and the Death Cap (see pp. 70, to pp. 74). This could never happen if the basic differences between the two genera were apparent to every mushroom collector; he should have it deeply imprinted on his memory that all Agaricus species have gills which are only initially white, but later turn pink, red, and finally purple-brown black. The gills of the Fool's Mushroom and of the Destroying Angel always remain pure white. Agaricus has no sheath at the base of the stem; it is always prominent on the Fool's Mushroom and on the Destroying Angel.—**Characteristics:** Cap: diameter up to 10cm, at first hemispherical, slightly convex when open, finally with the centre depressed. The cap has a dense cover of fine, brown, fibrous scales. Gills: at first rather pale but hardly pure white, gradually turning red, chocolate-coloured and finally almost black, which is an important characteristic. Stem: white, length up to 12cm, turns distinctly red when bruised and even when touched. Round the upper part it has a ring, which gradually darkens as spores collect on it. Flesh: white, but immediately turning red when cut, faintly smelling of freshly-sawn wood.—**Habitat:** Although its French name of champignon suggests that it grows in the fields, it is a typical woodland mushroom and thrives particularly in the litter of pine needles. Common in summer and autumn, growing in clusters and nests.—**Possibilities of confusion:** Some other Agaricus species resemble Agaricus silvaticus closely, but because they are also edible they can be collected without fear. Only in one case could confusion be unpleasant, that with Agaricus placomyces, which is poisonous; the base of its stem when cut turns brilliant chrome yellow. It can cause severe abdominal upsets.—**Use:** Good edible fungus.—**Relatives:** Agaricus langei Moell. is very similar to, but larger than, Agaricus silvaticus, and also edible. Another species, confined to deciduous woods, Agaricus haemorrhoidarius, turns very deep red. The agarics shown here enjoy greater popularity.

The Horse Mushroom

Agaricus arvensis

When squashed, injured or cut, mushrooms vary very widely in the discolouration of their flesh. The flesh of one group turns more or less red, that of the other distinctly yellow; the Horse Mushroom shows a yellow reaction.—**Characteristics:** Cap: diameter up to 20cm, bell-shaped at first, slightly convex later, initially pure white, where it had been touched and later generally yellow, silky sheen. Gills: at first grey, becoming red, flesh-coloured and finally dark chocolate-brown. Stem: white, length up to 15cm, slender, with a ring. Flesh: white, distinct smell of aniseed, an important mark of recognition.—**Habitat:** At the edge of woods, in clearings, under shrubs, in parks and gardens in summer and autumn.—**Possibilities of confusion:** Confusion can be dangerous with the Fool's Mushroom, the Destroying Angel (p. 70 top) and the Yellow-staining Mushroom, Agaricus xanthodermus, which is distinguished by its chrome-yellow flesh in the base of the stem.—**Use:** A delicious edible fungus, at least equal to the cultivated champignon.

The Field Mushroom

Agaricus campestris

When abundant rain has fallen after weeks of hot, dry weather that has brought the growth of fungi almost to a standstill is the time to look for the Field or Common Mushroom.—**Characteristics:** Cap: medium-sized, diameter hardly exceeding 10cm. At first hemispherical, later flattening, white, sometimes covered with fine, brown scales. Gills: at first pink, progressively darkening almost to black. Stem: average length 7cm, diameter up to 2cm, with a ring. Flesh: white, usually acquiring a very pale red hue, mild-tasting, but with no smell of aniseed.—**Habitat:** Often abundant in meadows, pastures and horse enclosures, in favourable weather.—**Possibilities of confusion:** In the neighbourhood of woods confusion with the Fool's Mushroom and the Destroying Angel (p. 70 top) is possible.—**Use:** A highly valued edible mushroom.—**Relatives:** About 30 different species are found in Britain, sometimes difficult to distinguish, or only with the aid of special literature. None are piosonous apart from the few species with distinctly yellow flesh in the base of the stem.

Inocybe patouillardii

Although the Fly Agaric and Inocybe patouillardii are of fundamentally different appearance, they contain the same poisonous substance, muscarine. But Inocybe patouillardii contains 20 times as much per unit weight as the Fly Agaric, which makes it far more dangerous.—**Characteristics:** Cap: diameter up to 8cm, conical, then opening up, with radial fibrils, often with a torn margin, at first white, turning brick red. Gills: white, later yellow to olive-coloured. Stem: length up to 7cm, diameter 1cm, at first white, later reddish. Flesh: white, also turning pale red.—**Habitat:** Appears early, in May and June, in meadows, at the edge of woods, under shrubs and in parks, in groups of many individuals.—**Possibilities of confusion:** Young specimens, which are still pure white, are easily confused with mushrooms or with the May Toadstool, Calocybe gambosa (p. 54 top), which appears at the same time. Older specimens are quite distinct because of their brick-red colour.—**Use:** A highly poisonous fungus which is found outside the main mushroom season.

Rozites caperatus

Rozites caperatus is one of those mushrooms which was rather neglected in the past. It has only gradually become known as a good eater which compares well with mushrooms bought in the market.—**Characteristics:** Cap: average diameter 10cm, hemispherical before opening up, with distinct boss. Straw-yellow and when young covered with a mauve-grey deposit resembling hoarfrost, often washed away by rain and hardly noticeable in old specimens, which often have a wrinkled cap. Gills: pale, becoming darker, with a light edge. Stem: length up to 12cm, diameter 2cm, fibrous with membranous ring. Flesh: white, with a pale violet sheen.—**Habitat:** Does not grow everywhere, but is common particularly on sandy soil in pine woods during summer and autumn.—**Possibilities of confusion:** Similar members of the genus Cortinarius never have a membranous ring.—**Use:** Rozites caperatus is an excellent edible fungus, but well known to and used by only a few collectors.

Cortinarius varius

Only those fungi in which collectors have always been interested have definitive names in the vernacular; others which are only gradually becoming established as edible have only Latin names until they are finally accepted and acquire a popular name. This has happened to Cortinarius varius in Germany where it is called 'Semmelfarbiger Schleimkopf'. In Britain it has no accepted common name.—**Characteristics:** Cap: diameter up to 10cm, at first hemispherical, later flat, the colour of a bread roll, yellow margin, slimy. Gills: a beautiful violet, only in old specimens discoloured brown by spore dust. Stem: whitish, club-shaped base, with the remnants of a web-like veil, which in young specimens joins the margin of the cap with the stem. Flesh: white, mild-tasting.—**Habitat:** Common in coniferous woods on chalky soil in autumn.—**Possibilities of confusion:** With similar members of the genus Cortinarius which, although not poisonous, may be inferior.—**Use:** Cortinarius varius, easily recognized owing to its blue-violet lamellae, is a good edible fungus which is still comparatively unknown.

Cortinarius violaceus

When this gorgeously coloured, striking fungus is found by one of a party of collectors it will be admired and passed from hand to hand. To fungus lovers who see it for the first time it is a memorable experience.—**Characteristics:** Cap: average diameter 10cm, at first hemispherical, with a slight boss when opened up. The scaly cap is violet, when old almost black-violet. Gills: violet when young, gradually turning brown from the deposit of spores, sinuate, with smooth edges. Stem: length up to 10cm, diameter up to 2cm, fibrous; bulbous base. The web-like veil between the margin of the cap and the stem is present in young specimens only. Flesh: grey-violet, smelling of cedarwood.—**Habitat:** In deciduous and coniferous woods, usually only a few individuals.—**Possibilities of confusion:** Its appearance is so characteristic that it is hardly ever confused with other fungi.—**Use:** Edible but rarely used.—**Relatives:** Cortinarius violaceus has recently been subdivided into more than one species.

Cortinarius praestans

The richly-developed pale-blue universal veil surrounds the stem like a boot. Young specimens, their deep-brown caps breaking through the white veil are said to resemble an owl's eye. The "praestans" of its Latin name indicates its unusual size.—**Characteristics:** Cap: hemispherical when young, sitting on a bulb, opening up to a diameter of up to 25cm, at first completely enclosed by the pale violet veil, from which the deep brown cap gradually frees itself; it becomes remarkably wrinkled later. Gills: whitish when young, narrow compared with the flesh of the cap, very crowded. Stem: length up to 15cm, diameter up to 5cm and even larger at the bulbous base, initially covered by the gradually disappearing, bluish, universal veil, finally white. Flesh: white, firm, hardly discolouring, mild taste, without distinct scent.—**Habitat:** Cortinarius praestans is not common, and usually found in small groups, probably confined to deciduous woods, near beech and oak trees, on chalky soil, scattered, and altogether absent from some regions.—**Possibilities of confusion:** Confusion with other large members of the genus Cortinarius growing in deciduous woods is an obvious possibility. But when the unusual size, the striking veil, and the wrinkled margin of the cap are considered the chances of confusion are remote.—**Use:** Cortinarius praestans is an excellent, and on account of its size, a very substantial edible fungus, highly prized in Switzerland and by French gourmets. In areas where it is rare the few locations where it does grow should not be revealed, and it should not be used in the kitchen.—**Relatives:** Cortinarius praestans belongs to the sub-genus Phlegmacium, represented in Europe by well over sixty species. Their identification calls for a detailed special study. The sub-genus does not include any poisonous fungi, but it does contain some very bitter oncs. Inexperienced mycologists are not advised to tackle this difficult group if they are not to lose their confidence and interest in fungi altogether.

Pholiota aurivella

As a rule, mushroom collectors equip themselves with a small basket and a knife. (A plastic bag is not recommended). But many a fungus lover would have liked to have brought a ladder or at least crampons along when he discovered brilliant yellow, very appetizing-looking fungi a few metres up on a tree trunk. Pholiota aurivella is one of the species growing high up on trees.—**Characteristics:** Cap: average diameter 12cm, in the centre yellow-brown, lighter along the margin. Very smeary during rain, with darker scales, which are sometimes washed off. Gills: sinuate, rather crowded, at first yellow but soon turning rust-brown from a deposit of spores. Stem: often much extended, horizontal in the decaying wood, curved upwards outside, yellow-brown, dry, length up to 10cm, diameter more than 1cm. Flesh: firm, yellow when young, darker towards the base of the stem, without distinct smell, but always quite bitter-tasting.—**Habitat:** Pholiota aurivella is not uncommon, but rarely found, because the fungus collector's eye is almost always glued to the floor of the wood to detect mushrooms among moss, herbs and grass. Not very often does he look up from the foot of the tree trunks to the branches, where Pholiota aurivella, usually too high to be within reach, erupts from the trunks of deciduous trees, especially the beech. It attacks both living and dying trees, and destroys them in the process. It is found in spring, April and May, and again in autumn, October and November.—**Possibilities of confusion:** A few similar members of the genera Pholiota and Gymnopilus exist, but none with a golden-yellow cap and growing so high above ground. Neither genus includes poisonous fungi, but some of them are too bitter to be palatable: Pholiota squarrosa, pale straw-yellow, with scaly stem and cap, growing in clusters at the foot of tree trunks is one of them.—**Use:** Pholiota aurivella, although it has a bitter taste, is sometimes described as edible; its flavour, however, is mediocre. On balance its use is not recommended.

Kuehneromyces mutabilis

There are several genera in the Agaric family which are particularly difficult to identify. They are fungi which take up a large volume of water in wet, and lose it in dry weather. This changes their appearance considerably, because when dry they are mostly very light-coloured, but become dark when moist; the change is continuous and leads the collector to believe that the fungi are all different. Fungi undergoing such changes are called hygrophanous; Kuehneromyces mutabilis exhibits this property most strikingly.—**Characteristics:** Cap: diameter up to 6cm, with a marked boss, cinnamon-coloured when moist, much paler when dry. The hygrophanous zone encircles the dry boss. Gills: at first pale, later also cinnamon-coloured, rather crowded and slightly decurrent. Stem: has a distinct, rather durable ring; below this it is squarrose (scaly); length up to 7cm, diameter hardly reaches 1cm. Flesh: pale, brown in the stem, indistinct scent and flavour.—**Habitat:** Grows only on wood; it much prefers stumps of deciduous trees and is found very rarely on those of conifers. As a very gregarious fungus it hardly ever grows as a solitary specimen, and covers entire stumps. Common throughout the Northern Hemisphere from spring to winter.—**Possibilities of confusion:** Not every agaric growing on wood is the genuine Kuehneromyces mutabilis. In the past, even the Honey Agaric (p. 58) was thought to be identical with it because of its similar appearance and habitat. Beware of confusing the fungus with the poisonous Sulphur Tuft, Hypholoma fasciculare (p. 96 bottom), which is easily distinguished by its green-yellow gills and bitter taste. In Kuehneromyces mutabilis the spore dust is rust-brown, in the Sulphur Tuft violet-brown.—**Use:** Kuehneromyces mutabilis is a popular edible fungus. Its gregarious presence in clusters allows effortless collection throughout the year. Particularly suitable for soups. It is one of the few kinds of fungus that can be cultivated on a suitable substrate, and for which the spawn is commercially available. The remarkable name of Kuehneromyces commemorates the distinguished Lyons mycologist Robert Kühner.

Hypholoma capnoides

The genus Hypholoma is represented in Britain by a dozen species, most of which are of scientific interest only. But two of them, Hypholoma capnoides and Hypholoma fasciculare, the Sulphur Tuft, are also of practical importance.—**Characteristics:** Cap: diameter up to 6cm, without scales or fibrils, at first convex, later flat, immediately after opening up with remnants of the veil along the margin of the cap, yellow-brown. Gills: at first pale, but soon turning smoke-grey, without yellow-green hues, which is a very important indicator. Stem: length up to 10cm, diameter hardly 1cm, upper part light in colour, dark brown towards the base, often curved, hollow. Flesh: white, perhaps a little astringent when chewed, but never really bitter.—**Habitat:** Always in clusters on conifer wood, common, appearing as early as spring but also in autumn, and can be collected during mild winters.—**Possibilities of confusion:** The poisonous Sulphur Tuft is distinguished by its yellow-green gills and bitter taste.—**Use:** A good edible fungus during months when fungi are scarce.

The Sulphur Tuft

Hypholoma fasciculare

The gills of the Sulphur Tuft are, as the name of the fungus implies, sulphur-coloured; this clearly distinguishes it from Hypholoma capnoides with its smoke-grey gills.—**Characteristics:** Cap: sulphur-yellow with brown centre. Gills: at first sulphur-yellow, later yellow-green, then olive-coloured, finally with a brownish deposit of spore dust. Stem: length up to 10cm, diameter hardly 1cm, yellow, with indistinct remnants of the veil. Flesh: sulphur-yellow and bitter.—**Habitat:** Common, in clusters on tree stumps, also during mild winters.—**Possibilities of confusion:** Frequent confusion with the edible Hypholoma capnoides; this is easily avoided by careful note of the colour of the gills.—**Use:** The fungus was regarded as inferior or unpalatable in the past; in fact it is poisonous. Collectors of agarics which grow in clusters on tree stumps must be able to identify it reliably.

The Verdigris Toadstool

Stropharia aeruginosa

A keen questioner once asked me whether a green variety of the Fly Agaric exists, claiming to have found it, with white flakes on the cap and a ring round the stem, exactly like the Fly Agaric except that the cap was green instead of red, at the bottom of his garden. He was told that he had not examined the gills carefully enough, because in the Fly Agaric they are white, but in the fungus he had found violet grey. To sum up: there is no green variety of the Fly Agaric; what he had found was the Verdigris Toadstool.—**Characteristics:** Cap: diameter up to 8cm, bell-shaped, with a boss, later flat, very slimy, verdigris-coloured with white scales floating in the slime. Old caps turn yellowish, and in dry weather are shiny. Remnants of the veil are often found on the margin of the cap. The skin can be peeled off. Gills: crowded, pale when young, then reddish grey, when old, violet-brown with white edges, sinuate. Stem: length up to 8cm, diameter 1cm, bluish green, with a membranous ring turning dark as it becomes covered with spores. Below the ring the hollow stem is fibrous and scaly. Flesh: white, slight smell and taste of radishes.—**Habitat:** Common in woods, meadows, on tree stumps and in gardens from summer to autumn.—**Possibilities of confusion:** The conspicuous colour prevents confusion. The white flakes on the cap may be washed off by heavy rain. A similarly coloured fungus, but of an altogether different shape, is Clitocybe odora, distinctly smelling of aniseed, with a funnel-shaped cap, without ring and with white spore dust.—**Use:** The Verdigris Toadstool was regarded as suspect in the past, but it is edible, although not held in high esteem. The slimy skin must be removed. Specimens growing by the roadside or near refuse dumps, a favourite habitat, should be rejected.—**Relatives:** A dozen representatives of the genus Stropharia are at home in Britain; they are, however, of no interest to the collector of edible fungi. In meadows, especially cattle pastures, Stropharia semiglobata may be found. As its name implies, its ochre cap is hemispherical. It grows on and very close to cow dung.

Anellaria semiovata

A large number of fungi grow on decaying matter, i.e. they are saprophytes living on rotting needles, leaves, wood, and other organic substances. Other fungi attack living organisms, such as trees, grass, even insects, i.e. they are typical parasites. Some species are not too choosey when it comes to selecting their nutrient substrate, others are strictly confined to certain media. Fungi, for instance of the genus Onygena, grow only on the hooves and horns cattle have shed, or on crows' feathers. Anellaria semiovata specializes in cattle dung.—**Characteristics:** Cap: at first egg-shaped, then hemispherical, finally bell-shaped; it never opens up like an umbrella; it is white, later cream-coloured, smeary when young, and showing cracks when dry. Gills: when the cap is cut it will be seen that the gills are adnexed, i.e. they reach the stem only at the very top and are hardly attached to it. They never autolyse into a black liquid, which is a typical feature of the genus Coprinus, but become patchy from the black spores, which do not all mature at the same time. The spore dust is deep black. Stem: strikingly long, up to 15cm, tapering from a thickened base, white, later beige, sand-coloured, hollow throughout its length. It carries a ring which is initially fluted, shrivels up very soon, is blackened by a deposit of spores, and often disappears altogether. Flesh: white, thin except in the centre of the cap.—**Habitat:** Anellaria semiovata is found from June to October, mainly on cattle pastures, where it produces its fruit bodies on cattle dung. Accordingly it rarely grows in woods or in arable areas, but prefers extensive cattle farming districts e.g. the mountain regions in Europe, especially Alpine pastures up to high altitudes. It is found throughout Europe.—**Possibilities of confusion:** A similar fungus is Strobaria semiglobata, which grows also on cattle dung, but is slimmer and has no ring even when young.—**Use:** Although Anellaria semiovata is not poisonous by any means, the environment it has chosen appears somewhat unappetizing and uninviting. Its value as an edible fungus has therefore not yet been established.

The Shaggy Ink Cap or Lawyer's Wig

Coprinus comatus

The Shaggy Ink Cap or Lawyer's Wig belongs to the genus Coprinus, which occupies a special position among the agarics. In this genus, too, the spores are of course produced by the gills but not all of them discharged as spore dust. When the spores are mature, cap and gills dissolve into an inky liquid, which contains many spores. The process of liquefaction is very rapid, and often takes only a few hours.—**Characteristics:** Cap: Height up to 12cm, when closed diameter up to 6cm, cylindrical with elliptical circumference. The fruit bodies often do not open up at all, but disintegrate whilst the cap is still closed. The cap is white, with brown hues in the centre, becomes red from the margin inwards when aging, and finally black. The surface of the cap soon turns shaggy, fibrous, which has earned it its American name "Shaggy Mane". But the most attractive appearance of the fresh cap lasts only a short while before the fungus collapses. Gills: unusually crowded, always free, white at first, turning first brown and quickly black, from the margin of the cap inwards and disintegrating into an inky liquid. Stem: length up to 15cm, diameter up to 2cm, surrounded by the closed cap for most of its length, hollow, does not become liquefied, and has a movable ring. Flesh: at first white, becoming reddish and finally black, pleasant scent and taste.—**Habitat:** Very common in the most varied types of meadow and lawn, particularly where manure has been used, in parks, but also on rubbish dumps, mostly in groups of many individuals. —**Possibilities of confusion:** The Common Ink Cap, Coprinus atramentarius is similar. It must not be confused with it because in certain conditions it may have a poisonous effect (see next plate).—**Use:** The Shaggy Ink Cap or Lawyer's Wig is an excellent edible fungus. Obviously only young specimens in which the gills are not yet discoloured i.e. still pure white, can be used. Because this fungus decays very quickly indeed, it should be used immediately after collection.

The Common Ink Cap

Coprinus atramentarius

If the collector is in contact with an expert on fungi, or joins a club of fungus lovers, identification of unknown mushrooms he has found on his excursions will present no problems. If he does not have these facilities, he can send doubtful specimens to experts by post. This is a quite satisfactory method, but Ink Caps, genus Coprinus, are quite unsuitable for it, because by the time they reach their destination, they will be a disgusting, black, liquid mess. The Common Ink Cap is one such fungus.—**Characteristics:** Cap: diameter up to 8cm, egg-shaped when young, later bell-shaped, distinctly pleated, so that the margin appears ribbed. As it opens up the cap tears from the margin inwards. It is grey, slightly yellow in the centre and covered with fine, tiny, mica-like scales, which can be easily wiped off. Gills: very crowded, often sticking together, white at first, but soon turning brown and finally black. Ultimately they dissolve into an inky-black liquid. Stem: average length 12cm, diameter more than 1cm, slender, later hollow, with a vestigial membranous ring and a distinct swelling at the base. Flesh: without notable scent and taste, whitish, later brownish.—**Habitat:** Very common, growing in clusters of many individuals and therefore conspicuous in meadows, parks, by the wayside, on rubbish dumps.—**Possibilities of confusion:** Must not be confused with the Shaggy Ink Cap or Lawyer's Wig, Coprinus comatus (see preceding plate).—**Use:** Young specimens with light-coloured, firm flesh, i.e. before they show symptoms of autolysis, may be eaten. It must, however, be borne in mind that alcoholic drinks must never be taken shortly before, during, and shortly after a meal of Common Ink Caps, because they are not compatible with this fungus; a weak poison that it probably contains normally passes through the alimentary canal without doing any harm but dissolves in alcohol, thus entering the bloodstream and causing illness.—**Relatives:** Coprinus picaceus is very pretty, with a cap covered with white flakes. The tiny Coprinus disseminatus grows in hundreds of individuals on old tree stumps; neither is edible.

The Sickener

Russula emetica

To be able to distinguish the hundred-odd species of Russula the collector needs years of detailed study of this genus. It is, however, relatively easy to determine whether an agaric is a Russula or not; the latter has white to yellowish gills which are almost invariably very brittle and break when touched. When a Russula drops onto a hard surface it breaks into pieces; other, softer, less rigid fungi lose only their stems. The reason for this difference lies in the structure of all members of the genus Russula: the fruit body consists not only of long-fibred cells, but also of many globular ones, the so-called spherocysts, which are embedded in it.—**Characteristics:** The Sickener is polymorphous as a species, i.e. it assumes so many shapes that it can be divided into several kinds or forms, which are adapted to various types of environment or biotope. Cap: diameter up to 10cm, brilliant bright red, margin grooved later, skin can be peeled off. Gills: white, sometimes with a very pale yellow tint; varying in length. Stem: white, length about 8cm, solid flesh, rigid and breaking easily. Flesh: white, smelling of fruit. A small sample is harmless but has a burning, peppery taste.—**Habitat:** Common in deciduous and coniferous riparian to Alpine forests in different versions, in summer and autumn.—**Possibilities of confusion:** The various red-capped Russulae are not at all easy to distinguish. But the following rule of thumb can be adopted for their examination: all Russulae which when sampled have a mild flavour are edible; all those whose flesh has an acid or bitter taste are inedible or poisonous. This rule of thumb is obviously valid only if the fungus to be tested has been positively identified as a Russula. The rule cannot be applied to other genera, where it could lead to the most dangerous confusion.—**Use:** The Sickener is a poisonous fungus, although slugs eat it with visible relish. No case of lethal poisoning by the Sickener has become known. It derives its name from causing vomiting.

The Green Agaric

Russula cyanoxantha

The Green Agaric should more accurately be called the Violet-green Agaric because of the colour of its cap; pure violet-blue specimens are rare.—**Characteristics:** Cap: average diameter 12cm, convex, later with a central depression, almost funnel-shaped when old. The skin can be peeled off the margin only. The colour varies from slate-grey to violet-green, which accounts for the name of the fungus; entirely green or very pale specimens are also found. Gills: provide a particularly important identifying mark. Unlike those of all other Russulae they are flexible and soft instead of brittle and very fragile. Stem: strong, length up to 10cm, diameter up to 2cm, white, sometimes with a red tint, and tapering towards the base. Flesh: white, without significant scent or taste.—**Habitat:** Common, much prefers beech woods, in summer and autumn.—**Possibilities of confusion:** Can be easily confused with very similar members of the genus Russula; this is, however, harmless because all mild-tasting species are edible.—**Use:** Highly valued edible fungus.

Russula vesca

Russula vesca is very popular among collectors.—**Characteristics:** Cap: diameter up to 10cm, slightly convex, later depressed, flesh-coloured. The skin does not reach the margin of the cap, so that the white ends of the gills form a narrow border. Gills: white, often with rust-coloured spots, spore dust pure white. Stem: length up to 7cm, tapering towards the base. Flesh: white, mild, nutty flavour.—**Habitat:** Common mainly in beech woods from the beginning of summer to autumn.—**Possibilities of confusion:** Similar mild-tasting species are edible. In cases of doubt it should be sampled.—**Use:** One of the most popular edible fungi. Although Russulae never become as soft as boletuses when cooked, this is a matter of taste, and may occasionally be preferred.

Russula ochroleuca

Certain fungi will not be found year after year, indeed some, because of their great rarity, will perhaps be seen only once in a lifetime. Russula ochroleuca is definitely not one of these. It grows in such profusion that it can fill whole laundry baskets almost every year.—**Characteristics:** Cap: diameter up to 9cm, up to half the skin can be peeled off, various shades of yellow, sometimes lemon-coloured. Gills: white, sinuate, becoming grey when older, length up to 8cm, diameter 2cm. Flesh: white, mild, acrid in the gills.—**Habitat:** Very common in coniferous woods in summer and autumn.—**Possibilities of confusion:** The bitter Russula fellea is very similar, but has darker gills and an intolerably acid taste, very easily detected when it is sampled. Grows in beech woods.—**Use:** Russula ochroleuca is edible, although it is not considered a first-quality table mushroom; but it is useful for mixing with other mushrooms.

Russula queletii

The collector of Russulae for the table should always make sure to place them in his basket with care and to avoid shocks as much as he can when carrying them home; if he does not take this precaution the fungi will break up, and he will have a rather discouraging mass of broken caps, gills, and stems in his basket by the time he arrives home. Russulae are, after all, very fragile fungi.—**Characteristics:** Cap: diameter up to 8cm, funnel-shaped when old, margin grooved, skin can be peeled off up to near the centre, purple with darker centre. Gills: whitish, never deep yellow, but turning green when bruised. Stem: length up to 7cm, diameter up to 2cm, red with bluish hues. Flesh: tastes very acrid.—**Habitat:** In pine woods during summer and autumn.—**Possibilities of confusion:** The intolerably acrid taste distinguishes it from all other red, mild and therefore edible Russulae.—**Use:** Inedible. After several acrid samplings the tongue has temporarily lost its ability to distinguish finer nuances of flavour.

Lactarius vellereus

The genus Lactarius closely resembles that of Russula. But unlike a Russula, a Lactarius, when injured, secretes a milky juice. The spherocysts mentioned earlier are found only in the flesh of Lactarius and never in the gills, which are therefore not brittle as in Russula:—**Characteristics:** Cap: as it opens up it often lifts part of the soil. Diameter up to 25cm, margin markedly involute, chalky-white, at first woolly to felt-like, funnel-shaped. Gills: whitish to ochre. Stem: white, stout, length up to 6cm. Flesh: very acrid taste.—**Habitat:** Common in deciduous and coniferous woods, especially in autumn.—**Possibilities of confusion:** Lactarius pergamenus and Lactarius piperatus are similar; both have crowded gills and bare caps, and can be eaten with meat courses when fried in fat; they never lose their bitter taste completely.—**Use:** Lactarius vellereus is unpalatable.

Lactarius turpis (Lactarius necator)

In old fungus books this drab-coloured fungus was called the Killer Agaric; the adjective of its Latin name means the same, but at worst it causes vomiting and diarrhoea, and its name appears to be needlessly alarming.—**Characteristics:** Cap: average diameter 12cm, with central depression, when old almost funnel-shaped, olive green. Gills: white, later yellowish, with dark spots. Stem: length up to 6cm, diameter up to 2cm, coloured like the cap but lighter. Flesh: white, profusely secreting a burning, acrid, milky juice.—**Habitat:** Very common, especially in pine woods, throughout Europe from summer into autumn.—**Possibilities of confusion:** The rule of thumb applied to Russula fungi according to which all mild-tasting species are edible, can be used also for Lactarius.—**Use:** Lactarius necator is not eaten in the West because it produces gastro-intestinal upsets. But it is widely collected in Eastern Europe, where it is a popular table mushroom after it has been rinsed, scalded, and the water poured off.

Lactarius volemus

A characteristic feature of the entire genus Lactarius is the secretion of a milky juice by all but old, dried-up specimens. The various species can be distinguished by the colour of their milk, which may be claret or carrot-coloured, yellow or violet, but is mostly white. The Lactarius volemus offers a rich supply of white milk.—**Characteristics:** Cap: average diameter 12cm, with involute margin, central depression, yellow to red brown, often cracked. Gills: light yellow, showing brown spots when bruised. Stem: usually a little paler than the cap, firm flesh. Flesh: yellowish-white, mild flavour, distinct smell of pickled herring.—**Habitat:** In groups of a few specimens, prefers beeches, in summer and autumn.—**Possibilities of confusion:** Similar species of Lactarius are usually less stout, and never have its characteristic smell.—**Use:** The Lactarius volemus is an exception in that it can be eaten raw. Like all the other species of Lactarius it is not suitable for boiling, which tends to make the flesh slimy, but should be fried.

Lactarius semisangifluus

The Lactarius semisangifluus secretes red milk.—**Characteristics:** Cap: diameter up to 12cm, at first depressed, later funnel-shaped, initially orange-red but soon turning green; old specimens are sometimes pure green. Gills: decurrent, bright orange-coloured at first, with green spots later. Stem: length up to 7cm, diameter up to 2cm, hollow. Flesh: white in the stem, marginal portions more or less reddish, when injured secreting a carrot-coloured milk, which turns blood red as it dries.—**Habitat:** abundant under young pine trees in Central Europe. It is doubtful if this occurs in Britain except perhaps as a chance import on roots of exotic conifers. Therefore there is no English name.—**Possibilities of confusion:** There are three other species of Milk Cap that secrete red milk; confusing them with this species is an advantage, because they are superior to it in flavour.—**Use:** Although the most inferior of the Milk Caps, this species is widely used in Europe in spite of its somewhat tart taste. After a dish of Red Milk Caps the urine turns red—no symptom of illness.

The Red Milk Cap

Lactarius rufus

Of the fifty-or-so species of Milk Cap at home in Britain about a dozen to varying degrees resemble the Red Milk Cap in the colour of their skin, in their milk, and in their general appearance.—**Characteristics:** Cap: average diameter 8cm, at first convex, but soon becoming depressed in the centre and developing a small pointed boss in the centre of the depression; this is a definitive distinguishing mark. The cap is evenly red-brown, the skin dry. Gills: pale reddish yellow, somewhat decurrent. Stem: average length 7cm, hollow, same colour as the cap but much paler. Flesh: secretes a very acrid, white milk.—**Habitat:** Abundant in coniferous woods, but only on acid soils, from summer to autumn.—**Possibilities of confusion:** Easily distinguished from similar species because of its pointed boss and the acid milk, which does not become discoloured.—**Use:** On account of its acrid taste not collected, but widely used in Eastern Europe after special treatment.

The Woolly Milk Cap

Lactarius torminosus

The Woolly Milk Cap belongs to the small group distinguished by a cap with a ragged fringe.—**Characteristics:** average diameter 10cm, margin at first strongly involute, then spreading and developing a funnel-shaped depression. The cap, particularly of young specimens, is distinctly felt-like, shaggy. Yellowish pink, with darker concentric bands. Gills: pinkish yellow, decurrent, mostly with a small tooth. Stem: length up to 6cm, pale pink, shallow-pitted. Flesh: faintly red, secreting a burning, acrid milky juice.—**Habitat:** Always associated with birch trees, therefore also found on moorland, throughout temperate Europe, in summer and autumn.—**Possibilities of confusion:** Its acrid, white milk is a feature clearly distinguishing it from the true, edible Milk Caps, with which on superficial examination it might be confused because of the reddish colour of its cap.—**Use:** The Woolly Milk Cap is poisonous, but like the Red Milk Cap used in Eastern Europe.

The Chantarelle

Cantarellus cibarius

The Chantarelle is perhaps the most popular and best-known of all the edible wild fungi. Possible reasons for this are that it is rarely attacked by grubs, easily identified, keeps relatively well, and can therefore be transported without damage.—**Characteristics:** Cap: diameter very rarely exceeding 5cm, giant specimens are, however, occasionally found; at first convex, later funnel-shaped, usually the colour of egg-yolk, but sometimes whitish or lilac. Ridges: ridge-shaped projections instead of agaric gills on the underside of the cap, very decurrent. Stem: like the whole fungus egg-yolk-coloured, tapering towards the base, upwards gradually merging into the cap. Flesh: yellowish white, strong peppery taste, which disappears during boiling.—**Habitat:** Very common except in districts where mushroom gathering is popular where it has already become quite rare.—**Possibilities of confusion:** The False Chantarelle, Hygrophoropsis aurantiaca, softer and more orange in colour, is not poisonous, but worthless all the same.—**Use:** Popular edible fungus, but containing much indigestible matter, unsuitable for drying.

The Horn of Plenty

Craterellus cornucopioides

This is a very tasty fungus well worth collecting.—**Characteristics:** The fruit body is not divided into cap and stem, but has the appearance of a funnel-shaped cornucopia, hence the Latin name. It grows to an average height of 10cm, opening towards the top from a narrow tube to the shape of a trumpet. The outside of the funnel is grey, smooth at first, pleated later, and produces the spores. The inside is scaly, in dry weather grey-brown, in wet weather almost black.—**Habitat:** The Horn of Plenty is found especially in beech woods, usually in clusters of many young and old individuals, crowded and therefore easy to collect, from the beginning of July to the end of the year.—**Possibilities of confusion:** There are no poisonous fungi resembling the Horn of Plenty.—**Use:** Young specimens, particularly when dried, are a tasty, highly valued seasoning for soups, gravies, and meat loaf.

Laetiporus sulphureus

Tubular fungi that grow on trees (bracket fungi) are generally of little interest to the fungus lover who collects for the kitchen unless he wants to use the fruit bodies as wall ornaments, because most bracket fungi consist of cork-like, woody, indigestible structures, often growing so high up on tree trunks that they are almost inaccessible. Laetiporus sulphureus is an exception. When young it is so highly prized that delicatessen shops in Europe may offer it as a special delicacy.—**Characteristics:** Cap: numerous caps develop one above the other, roof-tile fashion, inter-connected to form composite fruit bodies of up to 50cm width and 1m length, weighing many kilograms. At first the colourful caps are a brilliant sulphur yellow, later more orange-red. The gorgeous colours fade as the fungus ages. Tubes: also sulphur yellow, length about 3mm, with very fine, round pores. Stem: very short, the caps are mostly sessile (directly sitting) on the tree trunk. Flesh: when young yellow, soft, very juicy, somewhat acid-tasting, soon becoming pale, then dry, of cheese-like consistency, and friable. Finally all that remains of the gorgeous young fungus is a group of dirty-white brackets.—**Habitat:** Grows on various deciduous trees (though rarely on conifers), particularly likes the willow (Salix), but also fruit trees. It attacks living trees, i.e. it is a parasite; it unfailingly kills its host within a few years; it can therefore be repeatedly harvested from the same tree. Appears as early as May and is easy to spot even from a distance because of its brilliant yellow colour.—**Possibilities of confusion:** There is no other bracket fungus of both this size and brilliant yellow colour; confusion is therefore not likely. Old, bleached specimens are often mistaken for some other species.—**Use:** Only very young specimens of Laetiporus sulphureus are edible. It is highly prized by the gourmet who wants something exotic for a change to stimulate his palate. Scalding followed by crumbing is recommended. Older specimens are not tasty at all, but dry, friable, and useless.

The Wood Hedgehog or Urchin of the Woods

Hydnum repandum

The Wood Hedgehog is a spiny fungus as might be expected from its name, belonging to a family that has neither gills, tubes, nor pores. The spore-producing layer on the underside of the cap forms a dense field of spines resembling a stubble field. In spite of the fact that they share this characteristic the various kinds are often not closely related. The Wood Hedgehog is a typical representative.—**Characteristics:** Cap: diameter up to 10cm, often coalescing with neighbouring caps, colour of a bread roll, close resemblance to the Chantarelle. Spines: differing in length, but on average 5mm, pointed, yellow, and easily breaking. Stem: length up to 6cm, same colour but paler than that of the cap. Flesh: whitish, in young specimens acrid taste, older ones are bitter.—**Habitat:** Common in deciduous and coniferous woods in summer and autumn.—**Possibilities of confusion:** Young specimens are often confused with Chantarelles. At higher altitudes a Wood Hedgehog of a more reddish yellow colour, Hydnum rufescens is found, which is also edible.—**Use:** Young specimens are very tasty; maggots usually avoid them.

Sarcodon imbricatum

The scaly surface of the cap of Sarcodon imbricatum resembles the plumage of a hawk, and its spiny underside doeskin.—**Characteristics:** Cap: diameter up to more than 20cm, margin involute in young specimens, later pitted in the centre, with coarse brown scabs that cover the entire surface.—Spines: very crowded, length up to 10mm, at first whitish, later grey-brown, brittle, slightly decurrent. Stem: length up to 6cm, diameter up to 2cm, whitish to brownish. Flesh: whitish, turning grey-brown, aromatic scent, mild in young specimens, bitter in older ones.—**Habitat:** mainly in coniferous woods, often in long rows or rings, in autumn.—**Possibilities of confusion:** With Sarcodon scabrosum, which is inedible and has a very bitter taste; it is easily identified by the greenish base of its stem or when sampled.—**Use:** Young fungi only are suitable, and especially after drying as a flavouring for sauces and gravies.

Claviaradelphus pistillaris

The collector will be surprised when he finds Claviaradelphus pistillaris for the first time, because he expects to distinguish a cap and a stem in a fungus. This giant Fairy Club is therefore often described as an Edible Boletus of which only the stem has developed, but not the cap. This mistaken identity is explained by the fact that it is a Fairy Club, in which the hymenium is not confined to tubes, gills, or spines, but covers the entire club-shaped fruit body, i.e. it produces spores on the whole surface of the fungus.—**Characteristics:** The club-shaped, handsome fruit body grows to a height of more than 20cm, and the club attains a maximum diameter of 5cm. When young the flesh is firm in the club; when old it becomes spongy and widely varies in size and shape. The colour is yellowish, leather-like, turning faintly brown later.—**Habitat:** It is found most often in beech woods on chalky soil, in damp, shady places throughout Europe in autumn and winter. Very rare in some regions.—**Possibilities of confusion:** Clavaria-delphus truncatus is very similar, but its club has a flattened top.—**Use:** The Fairy Clubs mentioned usually have a slightly bitter taste and as edible fungi are held in low esteem. They are not poisonous.—**Relatives:** The genus Claviaradelphus includes some very interesting types of fungus. Clavariadelphus ligula is much smaller than Clavariadelphus pistillaris, growing to a height of about 8cm in the needle litter of pine forests gregariously in many hundreds of individuals; it is edible but of little value. Clavariadelphus fistulosa is found among fallen leaves; it grows as tall as Clavariadelphus pistillaris but remains very slender, no more than 6mm in diameter, hollow, tubular, inedible. Clavariadelphus juncea is hardly club-shaped at all. Its fruit body, up to 10cm long, is thin, almost thread-like, no more than 2mm in diameter; it grows in the leaf litter of beech and oak trees in late autumn and is of no importance as an edible fungus.

Vegetable Tripe, Cauliflower Fungus

Sparassis crispa

This striking fungus, which can attain a weight of several kilogramme, can be compared to a hen sitting on its cluster of eggs, or to a bath sponge; comparison with tripe, too, is apt.—**Characteristics:** A round, cauliflower-shaped structure grows from a fleshy stump to an average width of 30cm and height of 25cm. The stump branches out and the ends of the branches are flattened like leaves and wavy. The fungus is at first pale yellow, the ends of the branches turn brownish.—**Habitat:** Always at the base of old pine trunks, or on pine roots or stumps. In regions with extensive pine forests not uncommon, but hardly found other than in areas of pines.—**Possibilities of confusion:** Sparassis laminosa is very similar, but found only in the company of oak trees.—**Use:** Both are excellent, substantial eaters. They must be most carefully cleaned to remove the small animals hidden in the labyrinth of the branches. Very old specimens, wilting or in which the ends of the branches have turned brown, are tough and bitter, and useless.

Ramaria formosa

Ramaria formosa and related species are also known as Goat's Beards. Some of them are edible, others, such as the "stomach-ache fungus" Clavaria pallida, of inferior quality or even poisonous. When old, edible and obnoxious species look very much alike and are difficult to distinguish from each other.—**Characteristics:** A bulbous white stump up to 4cm in diameter sends out strong flesh-coloured branches, which branch out in turn, with lemon-yellow tips. But the white of the stump, the flesh colour of the branches, and the yellow of the tips become indistinct in the old specimens.—**Habitat:** In deciduous woods, where it prefers the beech.—**Possibilities of confusion:** Several very similar species of Clavaria are collected as edible fungi, but there are also poisonous ones which are difficult to distinguish.—**Use:** Because species of Ramaria are easily confused it is recommended that they should not be used for the table. Accounts differ, but some are definitely poisonous.

The Common Earth Ball

Scleroderma aurantium

When you come across a Common Earth Ball for the first time you will be struck by its similarity with a potato; both appearance and weight are the same. This hard puff ball is also often mistaken for a truffle, but its resemblance to this very expensive flavouring fungus for the gourmet's table is only superficial. The Common Earth Ball belongs to the Gastermycetes, the truffle to Tuberales.—**Characteristics:** Like all Gastermycetes the Common Earth Ball forms an enclosed fruit body, which produces the spores internally; these are liberated only as the body decays. The fungus looks like a medium-sized potato, is fairly hard and comparatively heavy. When a body is cut through, a wall of about 3mm thickness will be seen. The inside of the fungus is initially rather light-coloured, but never pure white. When the spores mature the inside turns first grey and finally almost black, and is traversed by many white fibrils; the wall tears open, releasing the spores, which are carried away, often forming a small "dust cloud", by the wind. When cut open the fungus emits a pungent, offensive smell.—**Habitat:** The Common Earth Ball is lime-hating and therefore not found everywhere. It grows only where the soil has been decalcified to a great depth, and also likes peat.—**Possibilities of confusion:** Scleroderma verrucosum is distinguished by a thinner wall.—**Use:** The Common Earth Ball is poisonous, the thin-walled variety useless. Common Earth Balls have been used to adulterate expensive dried truffles.— The bottom plate shows an interesting feature peculiar to the Common Earth Ball: it shows an attack by a parasite fungus, Xerocomus parasiticus, which feeds on and damages the Common Earth Ball so that it is unable to produce mature spores; the parasite is closely related to the Ceps, the Goat's Lip, and Xerocomus chrysenteron. Strangely enough this fungus, although it feeds on a poisonous species, is itself non-poisonous and safe to eat.

The Common Puff Ball

Lycoperdon perlatum (Lycoperdon gemmatum)

Even as children, mushroom lovers were fascinated by the ripe fruit bodies of the Common Puff Ball. It was fun to kick them like tiny footballs, when they released from their burst-open walls clouds of "smoke", although the children did not realise that they consisted of innumerable spores.—**Characteristics:** The fruit bodies have the shape of a bottle turned upside-down, and reach a height of 8 to 10cm, and at the most bulbous level a diameter of 5cm. The entire fruit body is covered by very fragile, tiny, easily wiped-off warts; when they drop off they leave the fungus bare, the initially white to yellow colour disappears, and the paper-thin wall becomes light to dark brown. An opening appears at the vertex, through which, released by a shock or a falling raindrop, the spores escape. The inside of the stem is filled with a spongy mass, the head contains the spore-producing tissue, which is at first pure white but soon turns yellow, olive-coloured, and pulpy before becoming dry and dust-like.—**Habitat:** The Common Puff Ball is one of the most frequently found soft Puff Balls in the most varied woodlands from riparian to High Alpine forests and accordingly develops strongly deviant forms throughout the Northern Hemisphere.—**Possibilities of confusion:** About a dozen species of Lycoperdon inhabit Britain; the commonest of them is Lycoperdon pyriforme, which grows gregariously in dense clusters on decaying tree stumps.—**Use:** The Common Puff Ball is edible. A rule of thumb can be established for this kind of fungus, too: all the fungi belonging to this family are edible provided they are pure white inside. The following piece of advice should be observed by the collector: when Common Puff Balls are added to other fungi in a basket, they shed their many tiny surface warts through friction, and the whole collection will look as if it had been carried in a bag of semolina that had not been properly emptied. Common Puff Balls should therefore be kept in a separate container, or at any rate apart from other fungi.

Anthurus archeri

There can be few fungus lovers who have come across this bizarre shape of mushroom which, by the way, is also a member of the Gastemycetes. It was unknown in Europe until quite recently, and is probably an immigrant from Australia. It may have been introduced on plant roots, but is now known along the Rhine valley and in a few places in England.—**Characteristics:** The young fungus forms an egg-shaped structure resembling a Puff Ball, of about 4cm diameter. As it matures the ball opens and several closely spaced branches grow upwards; they then separate, looking a little like a starfish. The 4 to 6 branches are raspberry-red and covered by a dark mass, which contains the spores and gives off an intolerable stench of carrion.—**Habitat** In meadows, pastures, and on the edge of woods.—**Possibilities of confusion:** Unique in shape, colour, and smell.—**Use:** Inedible fungi of such exotic origin and astonishing beauty deserves preservation.

The Earth Star

Geastrum triplex

The experienced fungus lover makes a point of occasionally clearing away the thick layer of leaf litter, when he will find what looks like a group of tulip bulbs which have inexplicably found their way to this unlikely spot. The mycologist will congratulate him on his discovery of young specimens of the Earth Star.—**Characteristics:** After the balls open, a triple structure forms: inside there is a puff-ball-like ball, which lies on a shallow "dish"; the outermost layer splits into several lobes, surrounding the inner parts in a star-shaped pattern of about 10cm diameter.—**Habitat:** This Earth Star is found in the leaf litter of deciduous woods in a warm climate, where it is quite common, but easily overlooked because its colour blends with its surroundings; cosmopolitan in all continents.—**Possibilities of confusion:** Almost two dozen species of Earth Star are found in Britain and Europe, none as large, and none with a "dish".—**Use:** All Earth Stars, marvels of the world of fungi, are objects of amazement and admiration and should be preserved.

The Stink Horn, or Devil's Horn

Phallus impudicus

No sooner have we entered a wood than our expert companion claims, although he had never been in this neighbourhood before, that there must be Stink Horns here. The penetrating stench of carrion, noticeable long before the fungus is discovered, has given it away.—**Characteristics:** When young the Stink Horn is subterranean, forms a body the size of a chicken egg to a fist. After the vertex of the egg has broken through the top soil it bursts and the stem and cap grow from it to a height of up to 20cm at a surprising speed. It is white, hollow, and has a spongy, cellular structure; it ends in a thimble-shaped cap covered with olive-green slime, which gives off an offensive smell of carrion. This attracts insects, mainly flies, but also dung beetles, which distribute the spores contained in the greenish slime. When the slime has dripped off the cap, a cellular, pitted cone is all that remains, the stem becomes limp and collapses.—**Habitat:** The Stink Horn is very common mainly in coniferous but also in deciduous woods from June to autumn, gregariously in small groups throughout the temperate zone.—**Possibilities of confusion:** Phallus hadriani is found in coastal sand dunes and in vineyards. It differs from the Stink Horn in that the skin of its "egg" is reddish violet. If you are fortunate enough to find a Stink Horn with a structure like a lace veil on its stem you may have made a remarkable discovery, Dictiophora duplicata, an immigrant from America. Such a rare fungus should always be taken to a botanical institution.—**Use:** Mature Stink Horns are unpalatable because of their smell. There are addicts who prepare the "eggs" like roast potatoes and praise them as a delicacy or chop them raw into salad, when they have a radishy flavour; few fungus lovers, however, can muster any enthusiasm for such an exotic dish.

The Common Morel

Morchella esculenta

If you want to collect morels, do not look for them where you usually find your mushrooms, in large deciduous and coniferous forests; they grow in riparian forests in sandy loam, provided it is chalky enough, under elms, ashes, and poplars.—**Characteristics:** Cap: varies greatly in size and shape, average height 7cm, width up to 5cm. The entire surface has a network pattern of longitudinal and transverse ridges, not unlike a honeycomb, the colour of yellow leather, sometimes more grey. Stem: length up to 6cm, surface texture like fine bran, hollow. Flesh: white, wax-like, brittle.—**Habitat:** On burnt ground and in other localities in April and May after warm rain; found in the same places but not always year after year.—**Possibilities of confusion:** Several sub-species, which are all edible, have been described. Confusion with the poisonous Gyromitra esculenta would be dangerous; but this fungus grows in sandy pine woods, hates lime, and has a dark-brown cap with brain-like convolutions. Although it can be eaten after it has been scalded and the water poured off, its use is to be decidedly discouraged.—**Use:** The Round Morel is a much sought-after, popular spring fungus, but because of the many hiding places of insects in its cap it must be very thoroughly cleaned.

Morchella elata

If the Common Morel is described as "Baroque", Morchella elata must be called "Gothic" in appearance.—**Characteristics:** Cap: taller, more slender, more pointed and darker than that of the Round Morel. The almost parallel, conspicuous longitudinal ridges are characteristic; they soon turn black. Stem: white, branny surface, hollow. Flesh: as in the Common Morel.—**Habitat:** similar to that of the Round Morel, but Morchella elata is much rarer and favours Scotland and Northern England, whereas the Common Morel prefers Southern and Eastern England.—**Possibilities of confusion:** Because no morel is poisonous, confusion is without practical significance.—**Use:** Morchella elata, too, is edible, but inferior to the Common Morel in flavour. Sliced across, the fungi, when prepared for the table, look like small toothed wheels and washers.

Caloscypha fulgens

Morchella, Gyromitra, Leotia, and Sarcoscypha—as well as this fungus—all belong to the class Ascomycetes or sac fungi, a term understood by few fungus lovers, who are unable to find anything in them that reminds them even remotely of a sac. The sacs, in which the spores are produced, measure only a few thousandths of a millimetre and can be seen only by those collectors who examine their finds under a microscope.—**Characteristics:** Cap: at first globular, becoming hemispherical to cup-shaped, with torn margin, diameter up to 4cm; brilliant yellow to orange-red inside, whitish outside, often with a blue-green hue. Stem: very short, the fungus is sometimes sessile. Flesh: wax-like, very fragile.—**Habitat:** In Alpine coniferous woods in spring at the time when the daffodil (Narcissus poeticus) is in bloom, often in long chains. Not found in Britain but Common in Central Europe.—**Possibilities of confusion:** The Orange-peel Fungus, Aleuria aurantia, often found in the gravel of woodland footpaths, is similar.—**Use:** Both fungi are edible, but because they have little flesh are rarely collected.

Loetia lubrica

The water requirements of fungi vary; most of them thrive only when there is plenty of water available. All the same, there are fungi that grow even in the desert and steppe. Leotia lubrica is one of those species that can exist only in permanently damp places.—**Characteristics:** Fruit body: its upper part is cap-shaped, without tubes or gills because it is a sac fungus. The small head is yellow to ochre. The fruit bodies vary widely in size and shape. The lower, stem-like part is of the same colour, and filled with slime. The gelatinous flesh is without distinct scent and flavour.—**Habitat:** In very shady, damp localities in moss, in clayey, loamy soils that prevent the rapid drainage of rain water.—**Possibilities of confusion:** Other species of Leotia are very rare indeed.—**Use:** Leotia lubrica is edible, but rarely collected.

Sarcoscypha coccinea

This gorgeous fungus, praised in Britain as the most beautiful of all cup fungi, makes the finder's heart beat faster.—**Characteristics:** Cup: at first mug-, then bowl- or dish-shaped, diameter up to 5cm. The yellowish margin later becomes wavy and tears. The inside of the cup is a brilliant vermilion, the outside whitish, flaky, with matted, transparent hairs. Stem: short, fine-matted. Flesh: wax-like, fragile.—**Habitat:** The fruit bodies unfold immediately after the snow melts. Because the time of this varies according to altitude, Sarcoscypha coccinea can be found from February until May, and in very mild regions even during winter. The fungi grow always in damp places on dead, fallen twigs overgrown with moss. In some areas rare, but often overlooked.—**Possibilities of confusion:** There are similar, smaller, red, non-poisonous cup fungi, but most of them are tiny.—**Use:** It should not be broadcast that Sarcoscypha coccinea is edible. Any true nature lover will refuse to destroy this beautiful fungus; he does not consider every mushroom in terms of a cheap source of food. Increased leisure during weekends, longer holidays, the possibility of reaching even distant woods conveniently by car, have had the effect of more and more "nature lovers" who do not deserve this distinction, becoming interested in fungi. They cannot identify them, do not want to make the effort of getting to know them, and indiscriminately grab everything they see, large and small, because the roomy boot of the car imposes no limits on the quantity they can collect. The rich harvest is not shown to experts in order to expand the finder's knowledge; all he wants is to have the edible, wholesome mushrooms sorted out for him. As a result, sometimes as much as 90% of the fungi shown, beautiful and rare ones like Sarcoscypha coccinea among them, are consigned to the dustbin. If the expert's advice and warnings remain unheeded and over-collection is repeated by many collectors over many weekends, a large number of species will be threatened with extinction. The beauty of Sarcoscypha coccinea should move us to preserve, admire, and appreciate our native fungi.

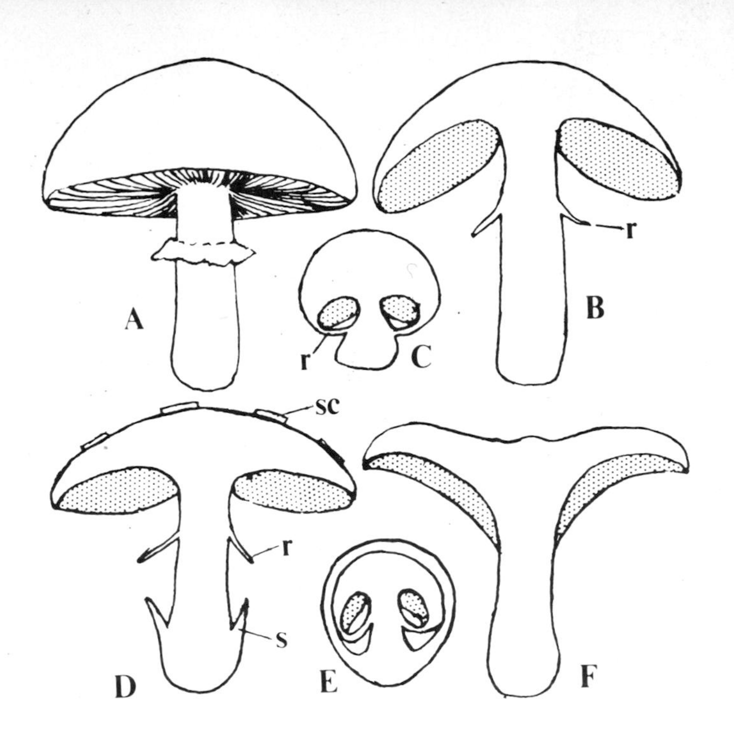

Mushrooms and Toadstools

A Mushrooms showing radiating gills on underside of cap.

B Section of mushroom (Agricus) showing adnate gills (stippled) and ring (r) on stem.

C Section of unexpanded mushroom showing origin of ring.

D Section of Amanita showing ring (r), scales on cap (sc) and sheath at base of stem (s).

E Section of unexpanded specimen showing universal veil (uv) which gives rise to sheath and scales, and partial veil, which gives rise to ring.

F Section of Russula with decurrent gills.

INDEX

English Names

Latin Names